Microscopic Examination
of the Activated
Sludge Process

WASTEWATER MICROBIOLOGY SERIES

Editor

Michael H. Gerardi

Microscopic Examination of the Activated Sludge Process

Michael H. Gerardi

Illustrations by Brittany Lytle

A JOHN WILEY & SONS, INC., PUBLICATION

Published by John Wiley & Sons, Inc., Hoboken, New Jersey
Published simultaneously in Canada

For general information on our other products and services or for technical support, please contact
our Customer Care Department within the United States at (800) 762-2974, outside the United States
at (317) 572-3993 or fax (317) 572-4002.

Wiley also publishes its books in a variety of electronic formats. Some content that appears in print
may not be available in electronic formats. For more information about Wiley products, visit our web
site at www.wiley.com.

ISBN: 978-0-470-05071-2

10 9 8 7 6 5 4 3 2 1

In Loving Memory of
John A. Gerardi, Jr.

The author extends his sincere appreciation to
Brittany Lytle for artwork used in this text.

Contents

Preface

Microscopic examinations of wastewater samples provide wastewater treatment plant operators with the opportunity to study "healthy" and "unhealthy" conditions in the activated sludge process and to decide on operational strategies for maintaining healthy conditions and preventing and correcting unhealthy conditions. Microscopic examinations also provide operators with immediate information regarding the condition of the biomass, response of the biomass to operational changes and industrial discharges, and the treatment efficiency of the activated sludge process.

There are numerous living and nonliving components in the mixed liquor of an activated sludge process that can be observed with a microscope and described with respect to desired or undesired numbers, structural characteristics, and activity. The observations and descriptions of these components provide indicators or "bioindicators" for process control and troubleshooting of the activated sludge process.

Significant living components of the activated sludge process include dispersed growth, floc particles, filamentous organisms, protozoa, rotifers, free-living nematodes, and zoogloeal growth. Additional living components include bloodworms (midge larvae), bristleworms, gastrotrich, spirochetes, tetrads, water bears, and water fleas. Significant nonliving components of the activated sludge process include colloids, gelatinous secretions, insoluble polysaccharides, and particulate material.

This book reviews the microscopic examinations of living and nonliving components of the mixed liquor as well as other wastewater samples including foam, scum, effluents, and recycle streams. The book also presents the use and care of the microscope and the stereoscopic binocular microscope, the rationale for the use of microscopy, and the procedures and techniques for performing a microscopic examination of the activated sludge process. Most importantly, the book presents the interpretation and correlation of observations with the health of the biomass and the efficiency of treatment.

Also presented in this book are worksheets, evaluation and rating tables, directions for preparing and applying microbiological stains and immobilizing agents, and techniques for preparing wet mounts and smears. The book is richly illustrated with photomicrographs and black-and-white drawings of many living, microscopic components of the activated sludge process.

Microscopic Examination of the Activated Sludge Process is the sixth book in the Wastewater Microbiology Series by John Wiley & Sons. The series is designed for wastewater personnel, and the series presents a microbiological review of the significant groups of organisms and their roles in wastewater treatment facilities.

MICHAEL H. GERARDI
Linden, Pennsylvania

Part I

Overview

1

Introduction

The microscope provides the wastewater treatment plant operator with a special tool for process control and troubleshooting of the activated sludge process. The microscope may be used on a routine or as-needed basis to determine the impact of various operational conditions on the biomass and the treatment efficiency of the activated sludge process. Sampling and frequency of microscopic examinations will be determined by manpower availability, severity of problems, and quantity and quality of industrial discharges. Frequencies from daily, weekly, and monthly to once every mean cell residence time (MCRT) have been used. However, during undesired operational conditions the microscope may be used more frequently to provide useful data with respect to the causative factors for the operational condition and progress of corrective measures.

Although the use of the microscope may be confusing at first and the amount of time spent on microscopic examinations considerable, the amount of time spent on microscopic examinations decreases greatly as microscopic techniques and the identification of organisms improve. Additionally, the operator need not be a microscopist to be able to use the microscope. Therefore, the use of the microscope may be incorporated as a standard analytical tool for process control and troubleshooting of the activated sludge process.

The microscope enables an operator to see the "bugs" or organisms in the treatment process. Each treatment process has its own profile of organisms when operating at a steady-state condition. By seeing the organisms the operator is able to correlate the organisms with existing operational conditions. These conditions may be acceptable or not acceptable. Therefore, the operator can "read" the organisms and determine whether operational conditions are acceptable or not acceptable, that is, use the organisms as indicators or "bioindicators."

Microscopic Examination of the Activated Sludge Process, by Michael H. Gerardi
Copyright © 2008 John Wiley & Sons, Inc.

The activated sludge process contains a large number and a large diversity of organisms. The major organisms by numbers and roles performed in the activated sludge process are the bacteria and protozoa (Figure 1.1). The minor organisms are the metazoa or multicellular, microscopic animals and macroscopic invertebrates (Figure 1.2). The metazoa commonly found in the activated sludge process include rotifers, free-living nematodes, water bears, and bristleworms. Additional organisms that are found in the activated sludge process include algae, fungi, immature insects, water fleas, and tubaflex.

Bacteria enter the activated sludge process through fecal waste and inflow and infiltration (I/I) as soil and water organisms and exist in the activated sludge process as dispersed and often motile cells, flocculated cells, and filamentous organisms. Dispersed cells include many young bacteria and the nitrifying bacteria *Nitrosomonas* and *Nitrobacter*. As bacteria age they lose the flagella that provide them with locomotion, and they produce a sticky polysaccharide coat that permits flocculation and floc particle development. *Escherichia* and *Zoogloea* are floc-forming bacteria, and quickly flocculate and initiate floc formation and development of floc particles. The dispersed cells that do not flocculate are (1) adsorbed to floc particles by compatible charge, (2) adsorbed to floc particles by change to compatible charge by the coating action of secretions from ciliated protozoa and metazoa, or (3) consumed or cropped by protozoa and metazoa.

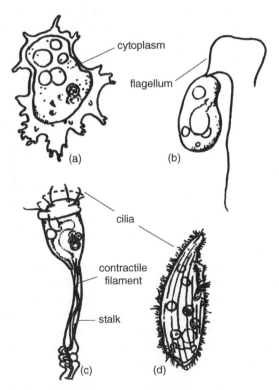

Figure 1.1   *Protozoa in the activated sludge process. There are three, commonly observed types of protozoa in the activated sludge process. These types include amoebae such as* Acanthamoeba *(a), flagellates such as* Bodo *(b), and ciliates such as* Vorticella *(c) and* Blepharisma *(d).*

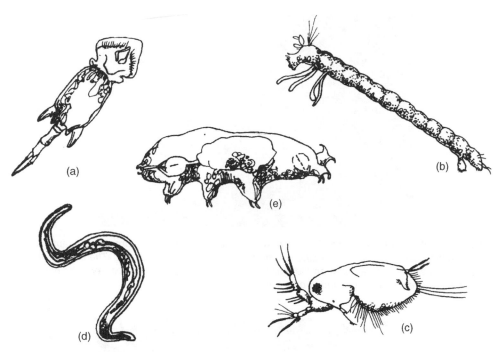

Figure 1.2 *Multicellular organisms in the activated sludge process. Several multicellular organisms commonly observed in the activated sludge process include the rotifer* (a), *the bloodworm* (b), *the water flea* (c), *the free-living nematode* (d), *and the water bear* (e).

Although they are single-celled organisms, some bacteria grow in chainlike fashion as trichomes or filaments. Large numbers of filamentous organisms enter activated sludge processes through three sources. They enter through I/I as soil and water organisms. They grow in the biofilm in sewer systems and enter through the sloughing action of the biofilm as wastewater flows over the biofilm, and they enter through the effluent of biologically, pretreated industrial wastewater. Examples of commonly occurring filamentous organisms in activated sludge processes include the Nocardioforms (Figure 1.3), *Microthrix parvicella* (Figure 1.4), and *Sphaerotilus natans* (Figure 1.5).

Bacteria are found in the activated sludge process in millions per milliliter of wastewater and billions per gram of solids. Bacteria are responsible for the degradation of organic wastes, nutrient (nitrogen and phosphorus) removal, floc formation, and removal of colloids, dispersed growth, particulate material, and heavy metals.

Protozoa are single-celled organisms and often are referred to as animallike or plantlike in composition. Most protozoa are free-living and enter the activated sludge process through inflow I/I as soil and water organisms. Depending on operational conditions protozoa are found in highly variable numbers from <100 per milliliter to tens of thousands per milliliter. Although wastewater protozoa have been placed in four, five, and six groups, wastewater protozoa most often are placed in five groups in the activated sludge process. These groups include the amoebae, flagellates, free-swimming ciliates, creeping or crawling ciliates, and stalked ciliates (Figure 1.6).

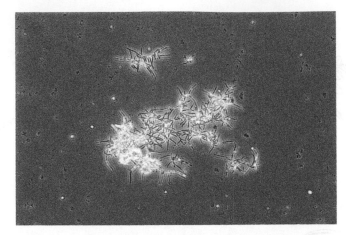

Figure 1.3   *Nocardioforms. Nocardioforms or Nocardia are short (<20 μm), branched, Gram-positive (blue) foam-producing filamentous organisms. Branching is true in Nocardioforms. There is a continuity of growth of cellular material between the branches, that is, there is no "gap" between the branches. Also, the branches are not surrounded by a transparent sheath.*

Figure 1.4   Microthrix parvicella. *Microthrix parvicella* is 100–400 μm in length and is a non-branched, Gram-positive (blue) foam-producing filamentous organism. The Gram-positive color of the filamentous organism often appears as a "chain of blue beads."

The gut content or cytoplasm of the amoebae is jellylike, and its cell membrane is very thin and flexible. The cytoplasm "flows" against the cell membrane and provides locomotion for the organism. The stretching and contracting motion of the amoebae is called a "false foot," or pseudopodia, mode of locomotion. This means of locomotion also enables the amoebae to trap particulate material and bacteria in order to obtain nourishment.

There are two types of amoebae, naked and testate. The naked amoebae such as *Amoeba proteus* (Figure 1.7) do not have a protective covering or testate. The testate amoebae such as *Difflugia* (Figure 1.8) do have a protective covering or testate. The testate consists of calcified material. It provides protection and allows the organism to drift in water currents. Amoebae move slowly through the wastewater and often

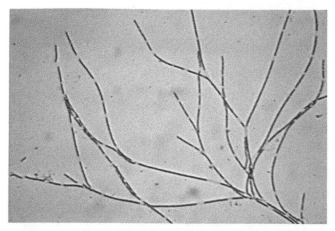

Figure 1.5 Sphaerotilus natans. Sphaerotilus natans *is a relatively long (>500µm), branched, Gram-negative (red) filamentous organism. Branching is false in Sphaerotilus natans. The branches are surrounded by a transparent sheath and have a "gap" or no cellular material between the branches.*

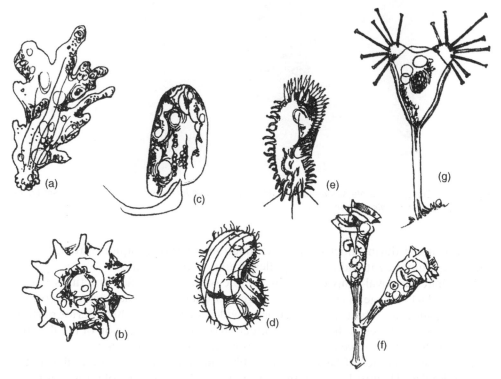

Figure 1.6 *Groups of protozoa. Representatives of the five basic groups of protozoa in the activated sludge process include the amoebae* Amoeba (a) *and* Arcella (b); *the flagellate* Cryptodifflugia (c); *the free-swimming ciliate* Colpoda (d); *the crawling ciliate* Styloncychia (e), *and the stalked ciliate* Epistylis (f). *Occasionally, a tentacled stalk ciliate such as* Acineta (g) *may be observed.*

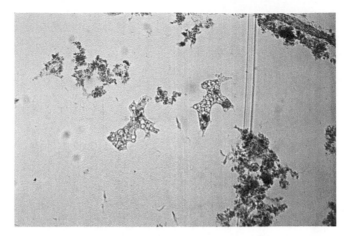

Figure 1.7   *Amoeba proteus. Naked amoebae such as* Amoeba proteus *often are difficult to observe because they move slowly and often have little contrast with their surrounding environment.*

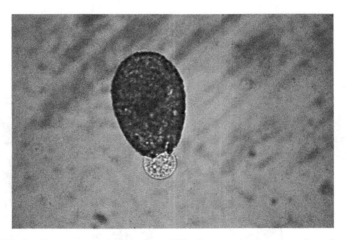

Figure 1.8   Difflugia. *Testate amoebae such as* Difflugia *move by the streaming of cytoplasm under the testate or drift in water currents.*

are overlooked during microscopic examinations of mixed liquor. Often, testate amoebae are misidentified as cysts, pollen grains, or other bodies.

Flagellates are oval-shaped and possess one or more whiplike structures, flagellum (singular) or flagella (plural). The flagella are located at the posterior portion of the organism, and their beating action provides locomotion. The anterior portion of the organism moves in the opposite direction of the beating action of the flagella. Therefore, flagellates have a quick, corkscrew pattern of locomotion.

There are two types of flagellates, plantlike and animallike. Plantlike flagellates such as *Euglena* (Figure 1.9) contain chloroplasts and are capable of photosynthesis. These flagellates also are known as motile algae. In the presence of excess phosphorus, pigmented flagellates may proliferate rapidly, resulting in population sizes >100,000/mL. Because pigmented flagellates are phototrophic, the swimming action toward the sunlight may cause bulking in secondary clarifiers when present in such

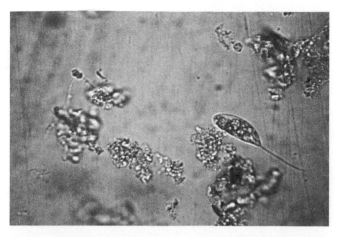

Figure 1.9 *Euglena. Plantlike motile algae or pigmented flagellates such as Euglena are green because of the presence of photosynthetic pigments or chloroplasts. These organisms are photo-trophic, that is, they swim toward sunlight. In addition to phototrophic ability, plantlike flagellates move by the whipping action of the flagella.*

Figure 1.10 *Swarming of pigmented flagellates. Relatively large numbers (>100,000 per mL) of pig-mented flagellates or motile algae are capable of "pushing" secondary solids toward the sunlight in outdoor settleability tests and clarifiers.*

large numbers (Figure 1.10). Animallike flagellates such as *Bodo* (Figure 1.11) do not contain chloroplasts.

Free-swimming ciliates swim freely in the bulk solution, that is, they do not attach to floc particles. Free-swimming ciliates such as *Paramecium* (Figure 1.12) and *Stentor* (Figure 1.13) possess numerous, short hairlike structures or cilia that are found in rows on the entire surface of the organism. The cilia beat in unison for locomotion and produce water currents for feeding purposes. The water current draws suspended or dispersed bacteria to the mouth opening on the bottom or ventral surface of the organism.

Crawling ciliates also possess cilia that are found in rows. However, the rows of cilia are located only on the ventral surface of the organism. Because of the reduced

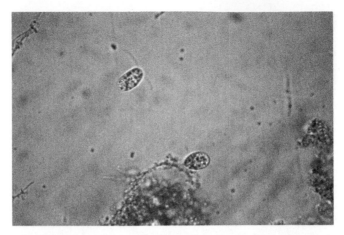

Figure 1.11 Bodo. *Animallike flagellates such as* Bodo *do not contain chloroplasts and are not phototrophic. Nonpigmented flagellates move exclusively by the whipping action of the flagella.*

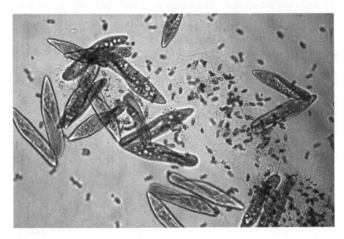

Figure 1.12 Paramecium. *Free-swimming ciliates such as* Paramecium *have rows of short, hairlike structures or cilia over the entire surface of the body. The beating action of the cilia provides locomotion and draws bacteria into the mouth opening.*

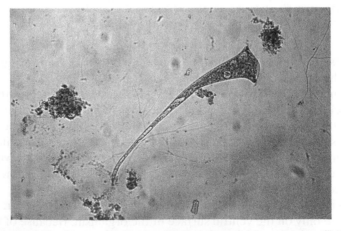

Figure 1.13 Stentor. Stentor *is a trumpet-shaped, free-swimming ciliate.*

number of cilia, crawling ciliates such as *Aspidisca* (Figure 1.14) and *Euplotes* (Figure 1.15) are poor swimmers and prefer to remain on the surface of floc particles. Some of the cilia at the anterior portion and/or posterior portion of the rows are modified to form "spikes," or cirri, that anchor the organism to the floc particles. Once anchored, the beating action of the cilia produces water currents that draw dispersed bacteria to the mouth opening on the ventral surface of the organism.

Stalked ciliates possess a circular row of cilia around the mouth opening. The cilia serve two purposes. First, they produce water currents to draw dispersed bacteria to the mouth opening. Second, they serve as a "propeller" that permits the organism to swim from a low dissolved oxygen concentration ($\leq 0.5\,\text{mg/L}$) to a high dissolved oxygen concentration (Figure 1.16).

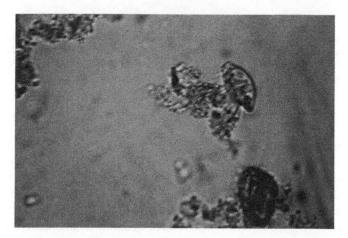

Figure 1.14 Aspidisca. *Crawling or creeping ciliates such as* Aspidisca *have rows of cilia only on the ventral surface of the body. The beating action of the cilia appears as numerous small "legs" as the protozoa "crawls" over the surface of a floc particle.*

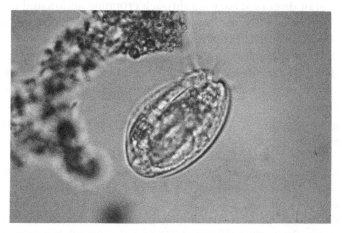

Figure 1.15 Euplotes. *Under high-power magnification, the cilia of* Euplotes *can be seen in contact with the surface of a floc particle. Some of the cilia have been modified to form "spikes," or cirri, that help anchor the protozoa to the surface of the floc particle.*

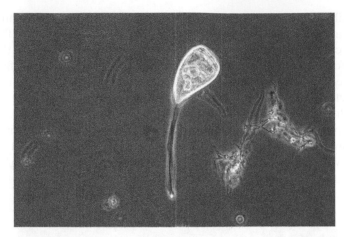

Figure 1.16 *Free-swimming stalked ciliate. Under low dissolved oxygen concentrations (<0.5 mg/L), stalked ciliates such as* Vorticella *detach from floc particles and swim towards higher dissolved oxygen concentrations. The stalked ciliates use their stalk or "tail" as a rudder and their cilia around the mouth opening as a "propeller."*

Stalked ciliates typically are sessile or attached to the floc particles but do swim freely under a low dissolved oxygen concentration. Stalked ciliates may be solitary (individual) such as *Vorticella* (Figure 1.17) or colonial such as *Carchesium* (Figure 1.18). Some stalked ciliates such as *Vorticella* possess a contractile filament that permits "springing" action (Figure 1.19). The springing action provides for a water vortex that draws more bacteria to the mouth opening. Some stalked ciliates such as *Opercularia* (Figure 1.20) do not have a contractile filament and cannot spring.

The protozoa, especially the ciliated protozoa, perform several significant and beneficial roles in the activated sludge process. These roles include:

- The removal of dispersed bacteria through cropping action and coating action. Cropping action is the consumption of bacteria, while coating action is the covering of the bacterial cell with secretions that make the surface charge of the bacterial cell compatible for adsorption to floc particles.
- Improved settling of floc particles (solids) in secondary clarifiers by adding weight to the floc particles when the protozoa are crawling on or attached to the floc particles
- Recycling mineral nutrients, especially nitrogen and phosphorus, through their excretions or waste products

Metazoa are multicellular organisms that enter the activated sludge process through I/I as soil and water animals. They are strict aerobes and do not tolerate adverse operational conditions such as low dissolved oxygen concentration, high pollution, and toxicity. The most commonly observed metazoa in the activated sludge process are the rotifer (Figure 1.21) and the free-living nematode (Figure 1.22). Although present in relatively small numbers (several hundred per milliliter), they too perform several significant and beneficial roles. These roles include:

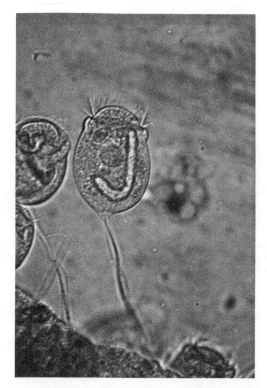

Figure 1.17   Vorticella. Vorticella *is a solitary protozoa.*

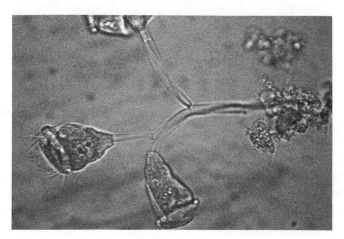

Figure 1.18   Carchesium. Carchesium *is a colonial protozoa.*

- The removal of dispersed bacteria through cropping action and coating action
- Improved settling of floc particles in secondary clarifiers by adding weight to the floc particles when they are crawling on floc particles or burrowing into floc particles

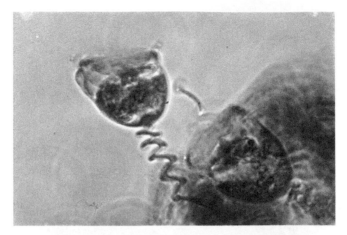

Figure 1.19 *Contractile filament. Some stalked ciliates such as* Vorticella *have a contractile filament or myoneme in the stalk that permits "springing" action. The springing action produces a water vortex that draws bacteria into the mouth opening.*

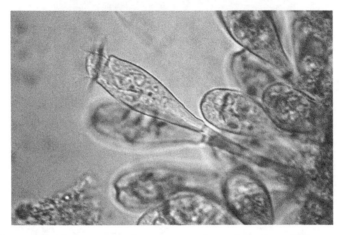

Figure 1.20 Opercularia. *Some stalked ciliates such as* Opercularia *do not have a contractile filament and are not capable of springing.*

- Recycling mineral nutrients, especially nitrogen and phosphorus, through their excretions or waste products
- The initiation of floc formation by secreting bundles of wastes that serve as attachment sites for bacteria
- Stimulating bacteria activity within floc particles by snipping away at floc particles or burrowing into floc particles. The snipping action and burrowing action permit free molecular oxygen (O_2), nitrate (NO_3^-), substrate or biochemical oxygen demand [(BOD)], and nutrients to penetrate to the core of the floc particles.

When the activated sludge process matures and operates successfully at a steady-state condition, the process has its own mixed liquor biota or "fingerprint" biota

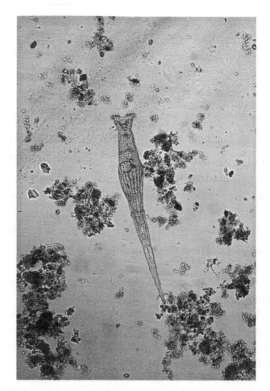

Figure 1.21   *Rotifer. Rotifers such as* Philodina *are the most commonly observed metazoa in the activated sludge process.*

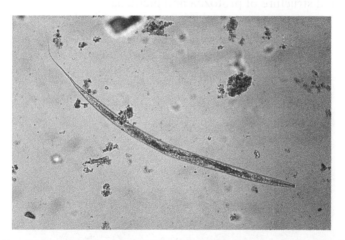

Figure 1.22   *Nematode. The free-living nematode is a commonly observed metazoa in the activated sludge and has mouthparts that are capable of biting into the floc particles.*

that reflects the steady-state environmental or operational condition of the process. This condition is noted in numerous parameters including (1) an acceptable range of values for the sludge volume index (SVI), (2) an acceptable range of values for the food-to-microorganism ratio (F/M), and (3) an acceptable range of values for the mean cell residence time (MCRT). At the steady-state condition the mixed liquor would consist of healthy life-forms that would be indicative of an acceptable mixed liquor effluent. For many activated sludge processes a mature and healthy mixed liquor biota under a steady-state condition (Figure 1.23) may contain the following:

- Little or insignificant dispersed growth in the bulk solution
- Little or insignificant particulate material in the bulk solution
- Mostly medium (150–500 μm) and large (>500 μm) floc particles
- Mostly irregular and golden-brown floc particles
- Mostly firm and dense floc particles as revealed through methylene blue staining
- Insignificant interfloc bridging and insignificant open floc formation
- Large and diverse population of ciliated protozoa

An undesired change in the mode of operation of the activated sludge process or an undesired change in industrial wastewater discharge can produce numerous changes in the mixed liquor biota that can be observed with microscopic examinations of the mixed liquor. Changes in the biota usually can be observed within 24–36 hours after the change in mode of operation or industrial discharge. Changes in the biota that can be observed include the following:

- Number of organisms
- Dominant and recessive groups of protozoa
- Activity and structure of protozoa and metazoa
- Quantity of dispersed growth and particulate material

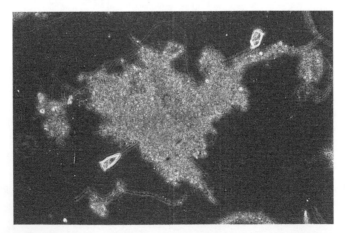

Figure 1.23   *Mature floc particle. In a healthy, steady-state condition a mature floc particle would be golden-brown in color, irregular in shape, and medium (150–500 μm) or large (>500 μm) in size. The floc particle would possess limited filamentous organisms growth. The bulk solution would contain little dispersed growth and little particulate material. Crawling ciliates and stalked ciliates may be found on the floc particle.*

TABLE 1.1   Reasons for Periodic Microscopic Examinations of Mixed Liquor

Correlate healthy and unhealthy biota to operational conditions
Correlate healthy and unhealthy biota to industrial discharges
Evaluate the impact of changes in the mode of operation
Evaluate the impact of industrial discharges
Identify factors responsible for loss of settleability
Identify factors responsible for loss of solids
Identify factors responsible for foam production
Identify appropriate process control measures
Monitor and regulate process control measures
Provide supportive data for industries and regulatory agencies

- Quantity of stored food in floc particles
- Strength and density of floc particles
- Dominant shape and size of floc particles
- Range in sizes of the floc particles

Periodic microscopic examinations of the mixed liquor can be used for many process control and troubleshooting purposes (Table 1.1). These purposes include in-plant and discharge monitoring and regulation. In-plant monitoring may simply be the use of target organisms or bioindicators to quickly determine deteriorating or improving conditions and appropriate process control measures. Discharge monitoring may be limited solely to the identification of problematic wastewaters.

2

Mixed Liquor Biota Food Chain

There is a large diversity of organisms that inhabit the mixed liquor and represent the biota of the activated sludge process (Table 2.1). These organisms enter the activated sludge process through fecal waste and I/I. Most organisms are free-living (nonpathogenic) and require the use of the microscope in order to be observed, that is, they are microscopic. A small number of organisms are macroscopic and can be seen without the use of a microscope. However, many macroscopic organisms are more easily observed with the use of the stereoscopic binocular microscope.

Development and maturation of the mixed liquor biota occurs through sludge aging, increasing dissolved oxygen concentration and decreasing pollution. Development and maturation may be described as a series of progressive steps, with each step having observable changes in the types, numbers, and dominance of organisms present and the quality of the bulk solution. These basic and progressive steps occur as follows with increasing sludge age or mean cell residence time (MCRT) and without operational upsets:

- Microscopic and macroscopic organisms enter the activated sludge process along with substrate or biochemical oxygen demand (BOD) and nutrients. BOD can be referred to as pollution. The organisms, substrates, and nutrients enter the process continuously through fecal waste and I/I.
- At start-up, the activated sludge process is heavily polluted and operates with a relatively low dissolved oxygen concentration. Lower life-forms such as bacteria, amoebae, and flagellated protozoa survive and increase in numbers under these conditions.
- With increasing sludge age and increasing numbers of bacteria, wastewater treatment efficiency improves. This permits an increase in dissolved oxygen

TABLE 2.1   **Mixed Liquor Biota**

Algae, filamentous	Fungi, unicellular
Algae, unicellular	Gastrotrich
Bacteria, dispersed	Nematodes, free-living
Bacteria, filamentous	Protozoa
Bacteria, floc-forming	Rotifers
Bloodworms (midge larvae)	Spirochetes
Bristleworms	Tetrads
Copepods	*Tubifex* (sludge worms)
Cyclops	Water bears (tardigrades)
Fungi, filamentous	Water fleas (daphnia)

concentration and a decrease in BOD. Lower life-forms continue to dominate the mixed liquor biota.

- As sludge age continues to increase, floc-forming bacteria undergo physiological stress and produce the necessary cellular components for agglutination and floc formation occurs. Floc formation results in the "packaging" of billions of organotrophic [carbonaceous BOD (cBOD)-removing] and nitrifying bacteria into small (<150 μm), spherical floc particles. The floc particles are white in color because of the lack of a significant accumulation of oils secreted by the bacteria. The development of floc particles results in continued improvement in treatment efficiency (increased dissolved oxygen concentration and decreased BOD). Bacteria in the mixed liquor are present in large numbers in the dispersed state and the flocculated state. The existing operational conditions permit the rapid growth of intermediate life-forms such as free-swimming ciliated protozoa. Ciliated protozoa help to cleanse the bulk solution of "fine" solids—colloids, dispersed bacteria, and particulate material.

- Filamentous organisms begin to increase in length and extend from the perimeter of the floc particles into the bulk solution. The increase in length of the filamentous organisms is caused by "stress" or sludge aging.

- The filamentous organisms help to reduce the cBOD of the wastewater and provide strength to the floc particle to overcome turbulence or shearing action. The filamentous network of strength enables the floc particle to increase in size as bacteria grow along the lengths of the filamentous organisms. The increase in size of the floc particle represents not only an increase in the number of bacteria but also an increase in the diversity of bacteria present in the mixed liquor. This permits even more efficient wastewater treatment of BOD as large numbers of slow-growing nitrifying bacteria oxidize ionized ammonia (NH_4^+) and nitrite (NO_2^-) to nitrate (NO_3^-) in the presence of high dissolved oxygen concentration.

- Because of the growth of filamentous organisms, the floc particles increase in size to medium (150–500 μm) and large (>500 μm). The floc particles also change to an irregular shape as the floc bacteria grow along the lengths of the filamentous organisms. With the increase in number and size of floc particles, more bacteria are found in the flocculated state rather than the dispersed state.

- With the shift in the majority of bacteria to the flocculated state, those protozoa with highly efficient feeding mechanisms for capturing the decreasing population of dispersed bacteria rapidly proliferate in the mixed liquor. The protozoa that proliferate are crawling ciliates and stalked ciliates. These protozoa attach to the floc particles and continue to clean the bulk solution of fine solids through coating action and cropping action.

- The floc particles become golden-brown in color. The change in color to golden-brown from white is caused by the accumulation of oils produced by old bacteria.

- In the environment of high dissolved oxygen and low BOD as provided by mature and stable mixed liquor, higher life-forms such as the metazoa (rotifers and free-living nematodes) are able to survive and are easily observed during microscopic examinations. However, the ability of metazoa to increase in numbers in the activated sludge process is severely hindered by (1) the long generation time of most metazoa as compared to the relatively short sludge age or MCRT of most activated sludge processes, (2) fluctuations in dissolved oxygen concentrations because of changes in wastewater strength and composition, and (3) the difficulty that female and male metazoa have in meeting and mating in the turbulent environment of the mixed liquor.

TRANSFER OF CARBON AND ENERGY: THE FOOD CHAIN

All organisms require carbon and energy for cellular activity and growth. Carbon and energy are available to the mixed liquor biota in the substrates found in the influent wastewater. The substrates are quickly absorbed and adsorbed by the organotrophic bacteria and the nitrifying bacteria in the wastewater and mixed liquor.

Organotrophic bacteria remove the carbonaceous BOD (cBOD) or organic compounds including acids, alcohols, amino acids, sugars, starches, lipids, and proteins. By degrading or oxidizing cBOD, the organotrophic bacteria obtain carbon and energy. Nitrifying bacteria obtain carbon from the alkalinity, primarily bicarbonate alkalinity (HCO_3^-) in the wastewater and obtain energy by degrading or oxidizing ionized ammonia (NH_4^+) or nitrogenous BOD (nBOD) in the wastewater. By oxidizing these compounds and removing alkalinity, organotrophic bacteria and nitrifying bacteria remain active and increase in number. The increase in number of bacteria is known as sludge production.

The bacteria in the mixed liquor also are cBOD, that is, living cBOD, because they serve as carbon and energy substrates for other organisms such the protozoa and metazoa that consume them. In turn, the protozoa and metazoa serve as living cBOD for organisms that consume them. In this fashion carbon and energy in nonliving cBOD are transformed to living cBOD and transferred through a food chain of organisms (Figure 2.1).

At a steady-state operational condition in an activated sludge process the mixed liquor biota can be characterized with respect to (1) dominant shapes and sizes of floc particles, (2) relative abundance and types of filamentous organisms, (3) quality of the bulk solution, (4) dominant protozoa, and (5) occurrence of metazoa. This biota should be seen consistently under microscopic examinations of the mixed liquor at steady-state operational conditions. However, changes in mode of operation and industrial discharges result in changes in the mixed liquor biota or food

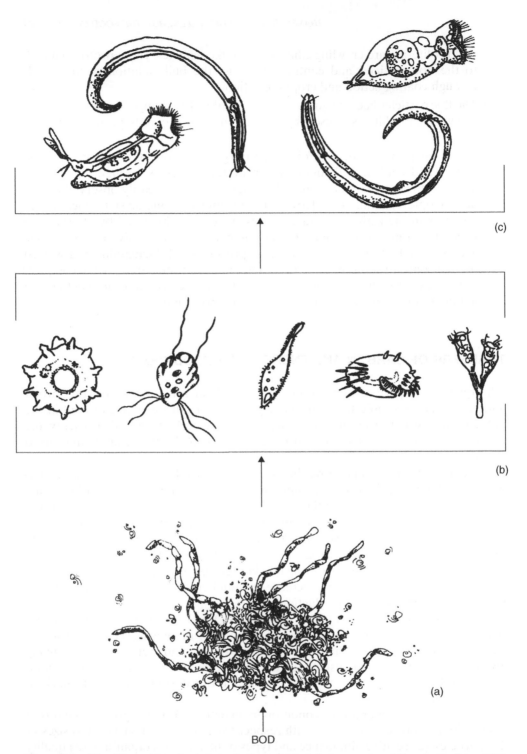

Figure 2.1 Mixed liquor biota food chain. In the activated sludge process, influent, non living BOD (substrate or carbon and energy supply) is degraded by bacteria for cellular growth or the production of new bacterial cells or sludge (a). The bacteria can be found in the dispersed or suspended state and the flocculated state as floc particles. By degrading the BOD, bacteria produce living BOD in the form of new bacteria. The bacteria in turn serve as substrate for the protozoa (b), and the protozoa in turn serve as substrate for the metazoa, rotifers, and free-living nematodes (c). In this manner, substrate is transferred through a food chain (bacteria to protozoa to metazoa) in the activated sludge process.

chain. These changes typically can be observed within 24–36 hours after a change in mode of operation or industrial discharge.

BACTERIA, THE MOST IMPORTANT GROUP OF ORGANISMS

Bacteria are the most important group of organisms in the mixed liquor biota or food chain. They are the most numerous and diverse group, and they perform the most significant roles in the activated sludge process. These roles include the following:

- Degradation of cBOD and nBOD
- Floc formation
- Removal of heavy metals
- Removal of nutrients (nitrogen and phosphorus)
- Removal of particulate material
- Removal of colloids

Bacteria are unicellular organisms, and most are <2μm in size. Most bacteria are found in three basic shapes: coccus or spherical, bacillus or rod-shaped, and spirillum or spiral (Figure 2.2). Other shapes such as rectangular, disk-shaped or barrel-shaped, and square are found. Bacteria may be found dispersed as single cells, clusters of cells (diplo- and tetrads), and chains of cells (trichomes or filaments) (Figure 2.3). Many bacteria are highly motile, especially young bacteria, and move by the beating action of whiplike structures or flagella (Figure 2.4).

In the activated sludge process bacteria are found in the dispersed state as unicellular organisms, clustered or agglutinated as floc particles, and in some conditions clustered as tetrads or groups of phosphorus-accumulating organisms (PAO) or poly-P bacteria and as filamentous organisms. The relative abundance of bacteria

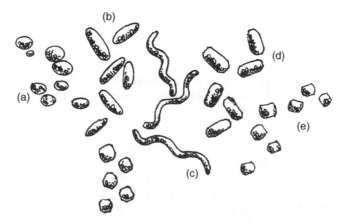

Figure 2.2 *Shapes of bacterial cells. There are three basic shapes for most bacterial cells. These shapes include coccus or spherical (a), bacillus or rod-shaped (b), and spirillum or spiral (c). Other shapes include rectangular (d) and disk-shaped (e).*

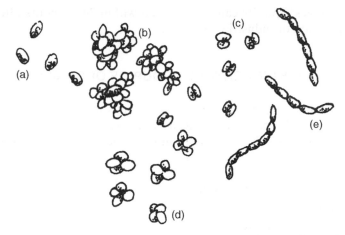

Figure 2.3   *Bacterial growth patterns. Bacteria may grow as individual or solitary cells (a) or together in odd-shaped clusters (b), pairs (c), tetrads or groups of four cells (d), and rows of cells or filamentous chains (e).*

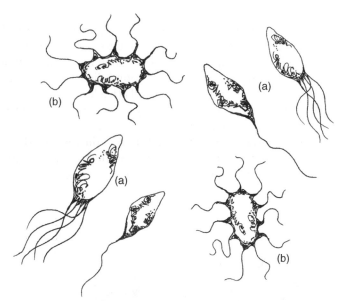

Figure 2.4   *Location of flagella on bacterial cells. Bacteria may possess one flagellum (whiplike structure) or two or more flagella. The flagella provide locomotion and are located on one end of the cell (a) or on the entire surface of the cell (b).*

in the activated sludge process is millions per milliliter of bulk solution and billions per gram of solids or floc particle.

There are numerous groups of bacteria in wastewater treatment plants that perform significant positive and negative roles (Table 2.2). Of these groups the most important are cBOD-removing bacteria, filamentous bacteria, floc-forming bacteria, nitrifying bacteria, and poly-P bacteria.

With respect to their response to free molecular oxygen (O_2), there are three groups of bacteria. These groups are the aerobes, the facultative anaerobes, and the

TABLE 2.2   Significant Groups of Wastewater Bacteria

Acetongenic bacteria	Hydrolytic bacteria
cBOD-removing bacteria	Methane-forming bacteria
Coliforms	Nitrifying bacteria
Denitrifying bacteria	Nocardioforms
Fecal coliforms	Pathogenic bacteria
Fermentative (acid-forming) bacteria	Poly-P bacteria
Filamentous bacteria	Saprophytic bacteria
Floc-forming bacteria	Sheathed bacteria
Gliding bacteria	Spirochetes
Gram-negative aerobic cocci and rods	Sulfur-oxidizing bacteria
Gram-negative facultative anaerobic rods	Sulfur-reducing bacteria

anaerobes. Aerobes are active only in the presence of free molecular oxygen. Nitrifying bacteria are strict aerobes. When they are present in relatively large numbers in the activated sludge process, they can be observed at 1000× total magnification as dense, rounded colonies at the perimeter of the floc particles.

Facultative anaerobes are active in the absence and presence of free molecular oxygen. They can use molecules such as nitrate (NO_3^-) but prefer free molecular oxygen. Denitrifying bacteria are facultative anaerobes. A genus of denitrifying bacteria that is easy to identify is *Hyphomicrobium*. This genus has a unique "bean on a stalk" structure. Anaerobes are inactive in the presence of free molecular oxygen. Some, such as methane-producing bacteria, die in the presence of free molecular oxygen. Besides methane-producing bacteria, the other significant anaerobic group found in wastewater treatment plants is the sulfate-reducing bacteria (SRB) that use sulfate (SO_4^{2-}) to degrade cBOD. Hydrogen sulfide (H_2S) is produced when sulfate is used to degrade cBOD.

3

Samples

Microscopic examinations of wastewater samples need not be limited to mixed liquor. There are numerous waste streams that can be sampled and examined microscopically for specific components that are of value in process control and troubleshooting of the activated sludge process (Table 3.1). Each waste stream may be examined for appropriate components such as (1) recycling of filamentous organisms, (2) contributing factors for foam production including filamentous organisms and nutrient-deficient floc particles, (3) characterization of floc particles, and (4) characterization of fine solids in the final effluent. An example of microscopic troubleshooting of a wastewater sample other than mixed liquor is the detection of translucent or transparent plastic resins or fibers in the final effluent that are revealed through methylene blue staining of a wet mount (Figure 3.1). The fibers are responsible for elevated total suspended solids (TSS). An additional example of troubleshooting of a wastewater sample other than mixed liquor is the detection of filamentous organisms growing over and choking the gills of fathead minnows (*Pimephales promelas*) that are used in whole effluent toxicity (WET) testing. Here, the filamentous organisms, not pollutants, would be responsible for the death of the minnows and the failure to pass a WET test.

Wastewater samples may be collected and examined immediately or refrigerated (4 °C, ±1 °C) in a plastic screw-capped bottle with an air space greater than the sample size for examination later. Most protozoan activity is inhibited at temperatures <4 °C.

Approximately 50 mL of sample is sufficient for microscopic examinations. Refrigerated samples should be allowed to warm to room temperature before microscopic examinations. Although a sample can be refrigerated for several days before examination, it should be examined within 48 hours of collection. Sample bottles should be labeled with the following information:

Microscopic Examination of the Activated Sludge Process, by Michael H. Gerardi
Copyright © 2008 John Wiley & Sons, Inc.

TABLE 3.1 Waste Streams Suitable for Microscopic Examination

Aerobic digester decant
Centrate and filtrate from sludge dewatering operations
Final effluent
Foam
Industrial discharges from biological pretreatment systems
Mixed liquor
Scum
Secondary clarifier effluent
Settleability test, after 30 minutes (foam/bubbles, floating solids, supernatant, and settled solids)
Thickener overflow

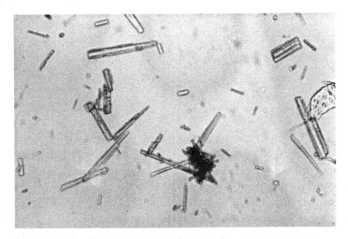

Figure 3.1 *Plastic fibers. Translucent or transparent, microscopic plastic fibers that contribute to elevated total suspended solids (TSS) in a final effluent can be easily observed with a microscopic scan of a methylene blue-stained wet mount of the effluent.*

- Name of the wastewater treatment plant
- Name and/or number of the treatment tank
- Type of sample (influent, effluent, recycle, scraping, overflow, foam, scum, mixed liquor, etc.)
- Date and time of sample
- Name and telephone number of the individual who collected the sample

The outside of the sample bottle should be rinsed thoroughly at the collection site, before the sample is taken to the laboratory for examination.

Mixed liquor samples should be collected from the aeration tank effluent. If an activated sludge process is operated with parallel systems, examine a mixed liquor sample from each system for possible indicators of an overloaded or imbalanced system. If foam occurs in the aeration tank, collect the mixed liquor without contamination from foam. If foam is to be examined, foam should not be contaminated with mixed liquor.

Any wastewater sample should not be diluted with water for microscopic examination. Where settling of solids has occurred, the wastewater sample should be

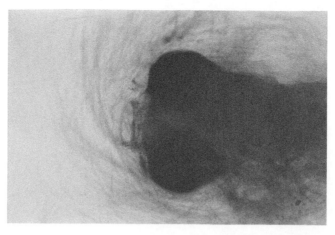

Figure 3.2 *Filamentous organisms type 021N growing on a fathead minnow,* Pimephales promelas, *under methylgreen stain.*

stirred to suspend the solids before an aliquot of sample is taken for microscopic examination. If protozoan analyses are to be performed, air should be forced along the sides and bottom of the sample bottle with a pipette and pipetting bulb to suspend amoebae before an aliquot of sample is taken for microscopic examination.

If wastewater samples are to be shipped to a laboratory for microscopic examinations, securely pack the bottle in an appropriate cooler with several freeze packs and ship the cooler for next-day delivery. Contact the laboratory before collecting and shipping the sample to ensure that the sample is collected, labeled, and shipped as required by the laboratory and that the laboratory can examine the sample as soon as possible after it is received.

Chemical preservatives should not be added to wastewater samples that are to be examined microscopically, and wastewater samples should not be frozen. Preservative and freezing adversely affect mixed liquor characteristics and the activity and structure of the mixed liquor biota.

Samples of mixed liquor should be collected and examined periodically, for example, once per week, to determine the significant components of the mixed liquor during a healthy, steady-state operational condition. The frequency of sampling and examination may be increased or decreased according to manpower availability, strength and composition of industrial discharges, changes in industrial discharges, and change in mode of operation. However, during adverse or upset conditions the frequency of sampling and the number of samples examined as well as the number of components examined per sample may need to be increased to identify the causative factors responsible for the upset and the appropriate corrective measures to be used (Figure 3.2).

4

Safety

Because wastewater samples do have or may have a large and diverse population of pathogens, there are several laboratory safety and housekeeping guidelines that should be followed when working with these samples to prevent or reduce the risk of infection. These guidelines include the following:

- Do not drink, eat, smoke, or chew gum or tobacco products in the laboratory.
- Do not remove any wastewater sample, wet mount, or smear from the laboratory.
- Use appropriate, protective latex or plastic gloves when working with waste-water samples, including wet mounts and smears.
- A clothespin may be used to hold a smear securely when performing a staining technique (Figure 4.1). Staining techniques should be performed over a sink.
- Thoroughly wash your hands with soap, warm water, and a disinfectant when you are finished with your microscopic examinations.
- Clean the laboratory counter with a disinfectant when you are finished with your microscopic examinations.
- Place all used glassware in a designated area for cleaning.
- Place paper towels, tissues, and gloves that have contacted any wastewater sample in a biohazard bag for disposal and not in the laboratory wastebasket.
- Only an appropriate worksheet for recording microscopic observations should be used at the workstation. Reference books should be stored and used at a site away from the workstation.
- Mouth pipetting of wastewater samples or reagents is not permitted.

Microscopic Examination of the Activated Sludge Process, by Michael H. Gerardi
Copyright © 2008 John Wiley & Sons, Inc.

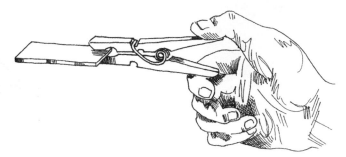

Figure 4.1   *Clothespin. A clothespin can be used to safely and securely hold a microscope slide for staining purposes. The use of the clothespin helps to keep stains off the fingers and hands.*

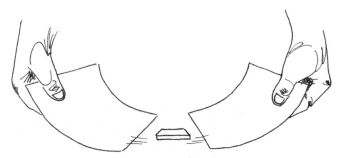

Figure 4.2   *Index cards. Dropped coverslips often are difficult to pick up. However, they can be safely removed from the floor or laboratory counter with the use of two index cards.*

- Immediately clean and disinfect any spills or accidents.
- Discard any used microscope slides and coverslips into a beaker with a disinfectant solution that contains a quaternary amine. Allow the slides and coverslips to soak in the disinfectant solution. The microscope slides and coverslips may be autoclaved. After disinfection, the slides and coverslips should be placed in a trash bag and disposed at a landfill. Do not recycle microscope slides or coverslips.
- If a coverslip should fall onto the laboratory counter or floor, two index cards may be used to pick up the coverslip (Figure 4.2). Attempting to pick up the coverslip with your fingers and thumb may result in a sliver of the coverslip cutting into and lodging in the skin.

Part II

Microscopy

5

Rationale for Microscopy

The organisms that are found in any habitat are there because the environmental conditions (dissolved oxygen, pH, substrates or BOD, temperature, other organisms, etc.) of that habitat favor their presence and proliferation. These environmental conditions also affect the activity and structure of the organisms. Conversely, the presence, relative abundance, activity, and structure of the organisms in a habitat reflect or indicate the relative quality (highly polluted, moderately polluted, or lightly polluted) and associated environmental conditions (dissolved oxygen, pH, substrates, temperature, toxicants, etc.) of the habitat. The organisms serve as indicators or bioindicators of the habitat. Because the activated sludge process is a habitat for a large and diverse population of organisms, these organisms can be used as bioindicators of the degree of pollution or health of the activated sludge process or mixed liquor. The relative abundance, dominance, and recessive groups of organisms, diversity of organisms, and their activity and structure can be easily observed with a microscope, that is, the microscope can be used as a diagnostic tool for process control and troubleshooting of the activated sludge process.

At a steady-state operational condition each activated sludge process has its own values or ranges of values for parameters of that operational condition including dissolved oxygen concentration, pH, substrates, temperature, toxicants, etc. The steady-state operational condition promotes the development of a steady-state, mixed liquor biota food chain (Figure 5.1) that can be described by the following microscopic observations:

- Bulk solution
  - Relative abundance of dispersed growth
  - Relative abundance of particulate material

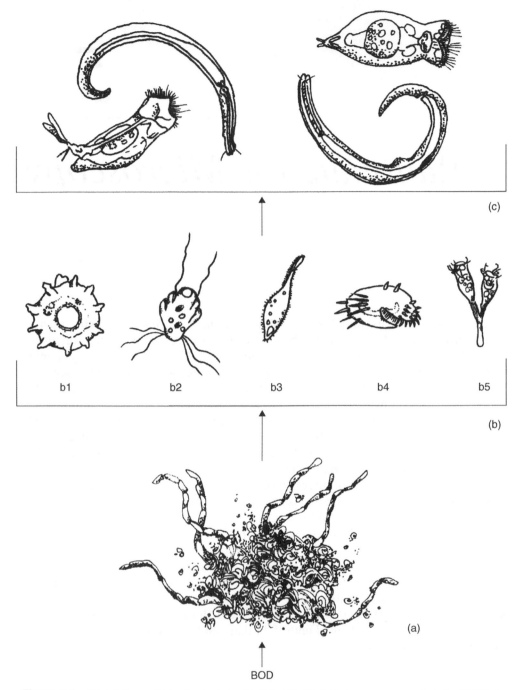

Figure 5.1 *Mixed liquor biota food chain. In the activated sludge process substrate (BOD) is degraded by bacterial cells. The degradation of BOD results in the growth of dispersed and flocculated bacteria (a). The bacteria serve as substrate for the protozoa (b) that in turn serve as substrate for the metazoa (rotifers and free-living nematodes) (c). There are five groups of protozoa in the food chain. The protozoa are amoebae (b1), flagellates (b2), free-swimming ciliates (b3), crawling ciliates (b4), and stalked ciliates (b5).*

- Floc particles
  - Range in size of floc particles
  - Dominant size of floc particles
  - Dominant shape of floc particles
  - Relative strength of floc particles
  - Response to the India ink reverse stain
  - Relative abundance of zoogloeal growth
- Filamentous organisms and floc particle structure
  - Relative abundance of interfloc bridging
  - Relative abundance of open floc formation
- Filamentous organisms
  - Relative abundance of filamentous organisms
  - Location of filamentous organisms
  - Dominant filamentous organisms
  - Recessive filamentous organisms
- Protozoa
  - Relative abundance of protozoa
  - Dominant protozoan groups
  - Recessive protozoan groups
  - Commonly observed genera or species of protozoa
  - Activity of protozoa
  - Structure of protozoa
- Metazoa (rotifers and free-living nematodes)
  - Relative abundance of metazoa
  - Activity of metazoa
  - Structure of metazoa
- Other higher life-forms (bristleworms, gastrotrich, etc.)
  - Higher life-forms observed
  - Activity of higher life-forms
  - Structure of higher life-forms
- Tetrads
  - Relative abundance of tetrads
  - Location of tetrads
- Spirochetes
  - Relative abundance of spirochetes

Significant and adverse changes in the steady-state operational condition of the activated sludge process result in significant and often easily observable changes in the mixed liquor biota food chain. Changes in the steady-state operational condition may be caused by (1) changes in wastewater strength and composition, usually industrial, (2) changes in mode of operation, and (3) changes in strength and composition of recycle streams.

Aside from inhibition and toxicity, changes in the steady-state mixed liquor biota food chain can occur within 24–36 hours after the operational change impacts or "hits" the aeration tank. Inhibitory and toxic conditions may impact the mixed liquor biota food chain immediately (30–60 minutes) for acute conditions or slowly (several days) for chronic conditions. Changes in operational conditions that have significant impact on the steady-state mixed liquor food chain are numerous (Figure 5.2) and include the following:

- Low dissolved oxygen concentration
- Low F/M (<0.05)
- Low nutrients (usually nitrogen or phosphorus)
- Low pH (<6.8)
- High pH (>7.5)
- Septicity/sulfides
- Shearing action
- Slug discharge of soluble cBOD
- Surfactants
- Toxicity, including heavy metals
- Fats, oils, and grease (FOG)
- Zoogloeal growth

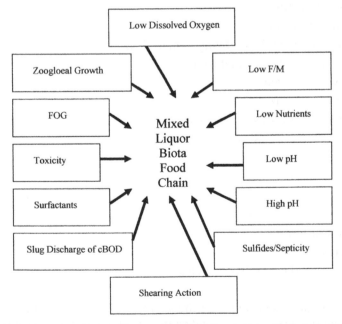

Figure 5.2   *Operational conditions affecting the mixed liquor biota food chain. Major operational conditions that affect the mixed liquor biota food chain include low dissolved oxygen, low F/M, low nutrients, low pH, high pH, sulfides/septicity, shearing action, slug discharge of soluble cBOD, surfactants, toxicity, FOG (fats, oils, and grease), and zoogloeal growth.*

Significant changes in the steady-state mixed liquor biota food chain because of adverse changes in the steady-state operational condition can be observed by the presence of undesired bioindicators, that is, changes in the following:

- Quality of the bulk solution
- Dominant size, shape, and strength of floc particles as well as range in size of floc particles
- Response of floc particles to the India ink reverse stain
- Relative abundance of zoogloeal growth
- Relative abundance of interfloc bridging and open floc formation
- Relative abundance of filamentous organisms, dominant and recessive filamentous organisms, and location of filamentous organisms
- Relative abundance of protozoa and dominant and recessive protozoan groups as well as commonly observed genera or species of protozoa
- Activity and structure of protozoa and metazoa
- Relative abundance of tetrads and spirochetes

There are specific and general bioindicators for each adverse operational condition. Some operational conditions have many bioindicators, while some operational conditions have only a few or one. The bioindicators for each adverse operational condition are provided below.

Low Dissolved Oxygen Concentration Bioindicators (Figure 5.3)

- Increase in dispersed growth
- Weak floc particles under methylene blue
- Increase relative abundance of filamentous organisms
  - *Haliscomenobacter hydrossis*
  - *Microthrix parvicella*
  - *Sphaerotilus natans*
  - Type 1701
- Regression of dominant protozoan groups from ciliates to flagellates and amoebae
- Sluggish activity or inactivity for protozoa and metazoa
- Numerous free-swimming, stalk ciliated protozoa

Low F/M (<0.05) Bioindicator (Figure 5.4)

- Increase in relative abundance of filamentous organisms
  - *Haliscomenobacter hydrossis*
  - *Microthrix parvicella*
  - Nocardioforms
  - Type 021N
  - Type 0041

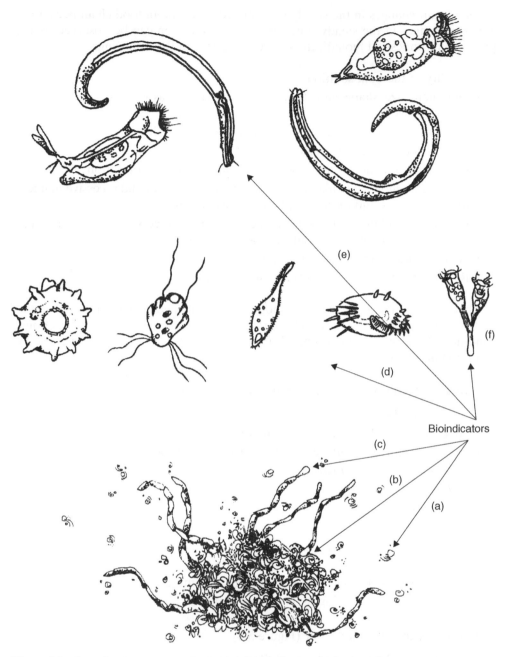

Figure 5.3 *Low dissolved oxygen concentration bioindicators. Indicators of a low dissolved oxygen concentration include increase in dispersed growth (a), presence of weak floc particles under methylene blue (b), increase in the number of filamentous organisms that grow under a low dissolved oxygen concentration (c), regression in dominant protozoan groups (d), sluggish activity of metazoa (e), and presence of numerous free-swimming stalked ciliates.*

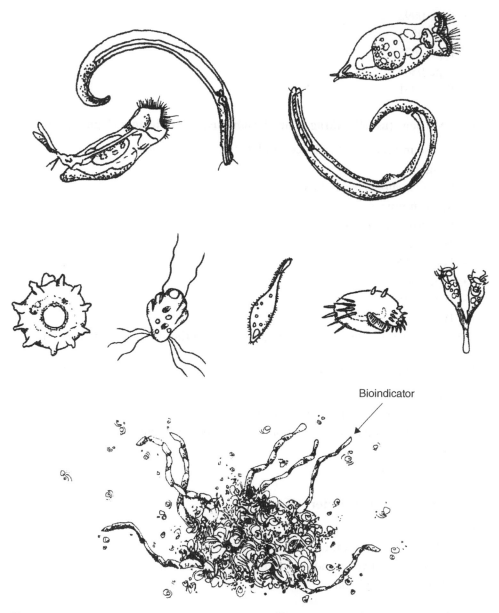

Figure 5.4   *Low F/M bioindicator. An indicator of a low F/M is an increase in the number of filamentous organisms that grow under a low F/M.*

- Type 0092
- Type 0581
- Type 0675
- Type 0803
- Type 0961

Low Nutrient (usually nitrogen or phosphorus) Bioindicators (Figure 5.5)

- Increase in relative abundance of filamentous organisms
  - Fungi
  - *Haliscomenobacter hydrossis*
  - Nocardioforms
  - *Sphaerotilus natans*
  - *Thiothrix* spp.
  - Type 021N
  - Type 0041
  - Type 0675
  - Type 1701
- Positive response by numerous floc particles to the India ink reverse stain

Low pH (<6.8) Bioindicators (Figure 5.6)

- Increase in dispersed growth
- Weak floc particles under methylene blue
- Increase in relative abundance of filamentous organisms
  - Fungi
  - Nocardioforms

High pH (>7.5) Bioindicators (Figure 5.7)

- Increase in dispersed growth
- Weak floc particles under methylene blue
- Increase in relative abundance of filamentous organisms
  - *Microthrix parvicella*

Septicity/Sulfides Bioindicators (Figure 5.8)

- Increase in dispersed growth
- Weak floc particles under methylene blue
- Increase in relative abundance of filamentous organisms
  - *Beggiatoa* spp.
  - *Nostocoida limicola*
  - *Thiothrix* spp.
  - Type 021N
  - Type 0041

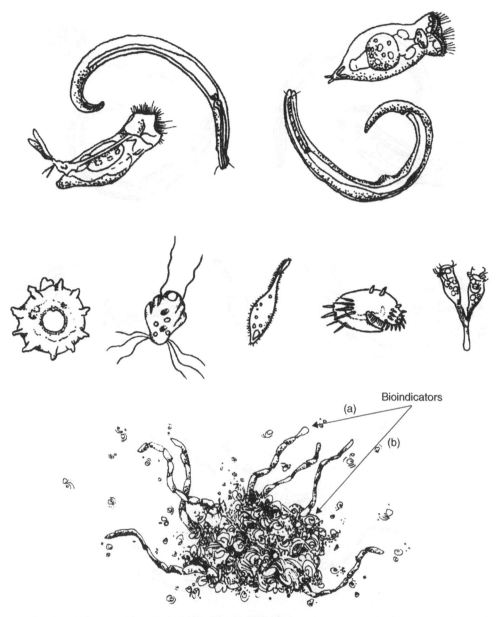

Figure 5.5 *Low nutrient bioindicators. Indicators of low nutrients (usually nitrogen or phosphorus) include increase in number of filamentous organisms that grow under a nutrient deficiency (a) and numerous positive responses of floc particles to the India ink reverse stain (b).*

Shearing Action Bioindicators (Figure 5.9)

- Increase in dispersed growth
- Presence of numerous small and irregular floc particles
- Presence of numerous free-floating and short filamentous organisms
- Presence of numerous sheared, stalk ciliated protozoa

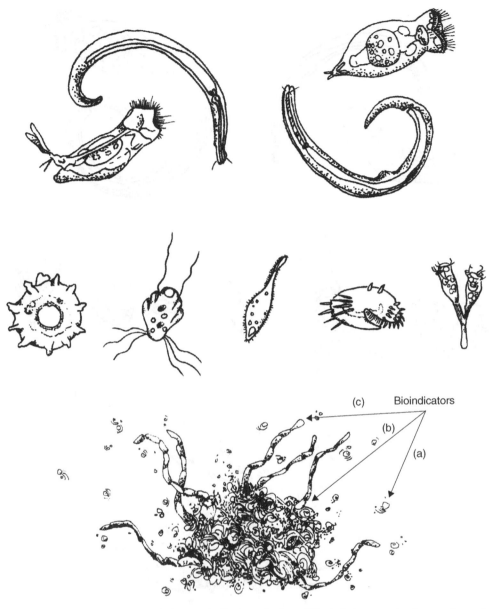

Figure 5.6 Low pH (<6.8) bioindicators. Indicators of a low pH include increase in dispersed growth (a), presence of weak floc particles under methylene blue (b), and increase in the number of filamentous organisms that grow under a low pH (c).

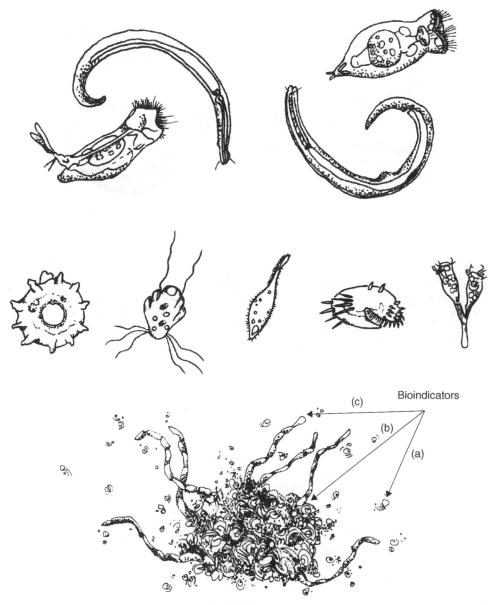

Figure 5.7 High pH (>7.5) bioindicators. Indicators of a high pH include increase in dispersed growth (a), presence of weak floc particles under methylene blue (b), and increase in the number of filamentous organisms that grow under a high pH (c).

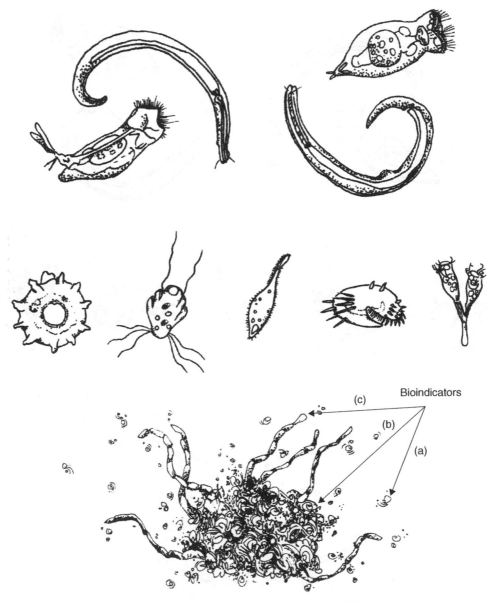

Figure 5.8 *Septicity/sulfides bioindicators. Indicators of septicity/sulfides include increase in dispersed growth (a), presence of weak floc particles under methylene blue (b), and increase in the number of filamentous organisms that grow under septicity/sulfides (c).*

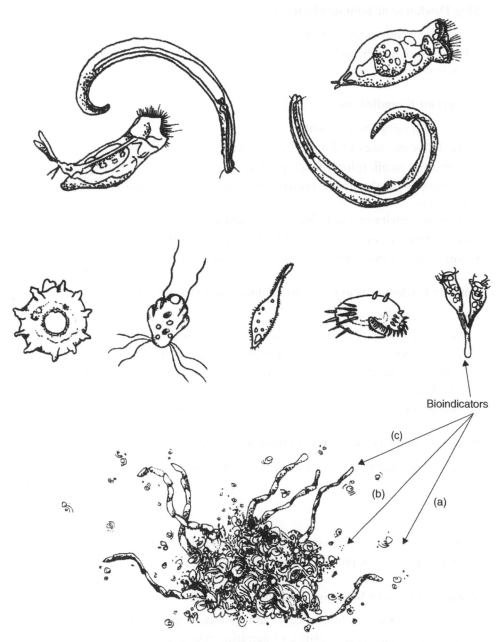

Figure 5.9   *Shearing action bioindicators. Indicators of shearing action include increase in dispersed growth (a), presence of numerous small and irregular floc particles (b), presence of numerous free-floating and short filamentous organisms (c), and presence of numerous sheared stalked ciliates.*

Slug Discharge of Soluble cBOD Bioindicators (Figure 5.10)

- Increase in dispersed growth
- Presence of numerous floc particles with a firm core and weak perimeter under a safranin stain

Surfactant Bioindicators (Figure 5.11)

- Increase in dispersed growth
- Weak floc particles under methylene blue
- Numerous small, spherical floc particles
- Regression in dominant protozoan groups from ciliates to flagellates and amoebae
- Sluggish activity or inactivity for protozoa and metazoa
- Dispersion of "soft" cells beneath the lorica of rotifers
- Dispersion of the cuticle in free-living nematodes

Toxicity (including heavy metals) Bioindicators (Figure 5.12)

- Increase in dispersed growth
- Weak floc particles under methylene blue, except for heavy metal toxicity
- Oval floc particles in presence of excess heavy metals
- Regression in dominant protozoan groups from ciliates to flagellates to amoebae
- Sluggish activity or inactivity for protozoa and metazoa

Fats, Oils, and Grease (FOG) Bioindicators (Figure 5.13)

- Increase in relative abundance of filamentous organisms
  - *Microthrix parvicella*
  - Nocardioforms
  - Type 0092
- Oil droplets under India ink reverse stain (Figure 5.14)
- Oil coating on floc particles (Figure 5.15)

Zoogloeal Growth Bioindicators (Figure 5.16)

- Increase in relative abundance of amorphous zoogloeal growth
- Increase in relative abundance of dendritic zoogloeal growth

A correlation of operational conditions and microscopic observations is presented in Table 5.1. The table may be used in two ways for troubleshooting the activated sludge process. First, a suspect operational condition such as low dissolved oxygen concentration may be considered highly probable or confirmed by determining the number of microscopic observations for the condition marked as "X" in the column for the condition during microscopic examinations of wet mounts and

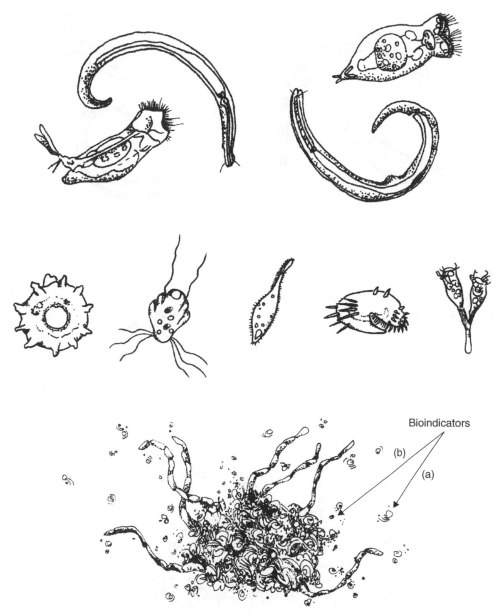

Figure 5.10   *Slug discharge of soluble cBOD bioindicators. Indicators of a slug discharge of soluble cBOD include increase in dispersed growth (a) and presence of numerous floc particles with a firm core and a weak perimeter under a safranin stain (b).*

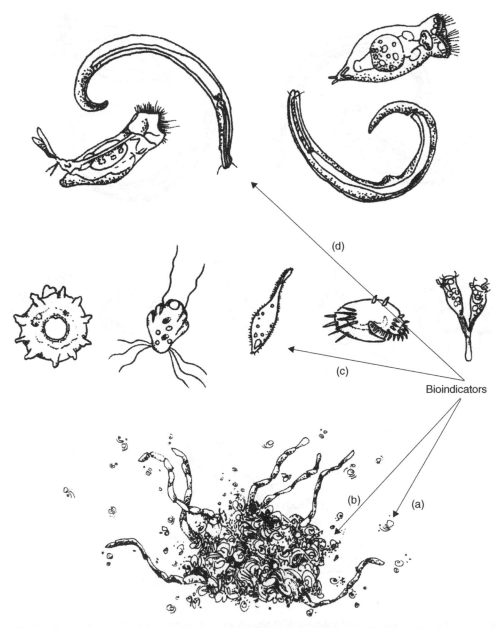

Figure 5.11 *Surfactant bioindicators. Indicators of excess surfactants include increase in dispersed growth (a), presence of weak floc particles under methylene blue (b), presence of numerous small, spherical floc particles (c), regression in dominant groups of protozoa and sluggish activity for protozoa (d), sluggish activity of rotifers and free-living nematodes (d), and dispersion of "soft" cells beneath the lorica of rotifers and the cuticle of free-living nematodes (d).*

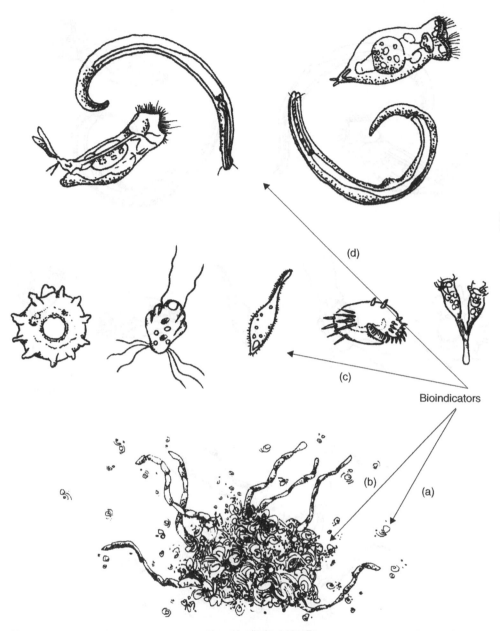

Figure 5.12   *Toxicity bioindicators. Indicators of toxicity include increase in dispersed growth* (a), *presence of weak floc particles under methylene blue* (b) *(except for heavy metal toxicity), regression of dominant protozoan groups and sluggish activity for protozoa* (c), *and sluggish activity for metazoa* (d).

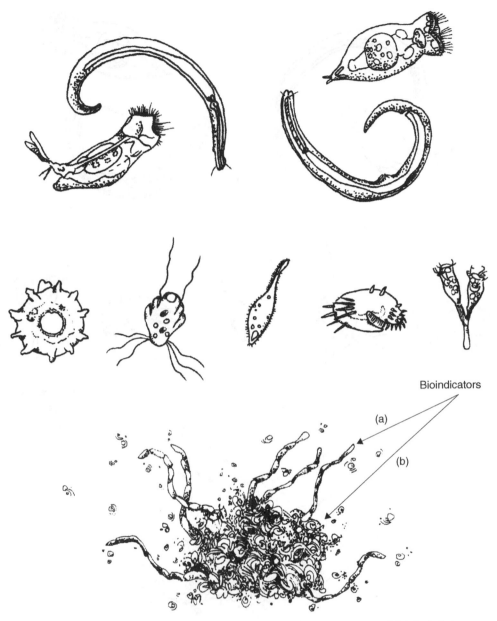

Figure 5.13   Fats, oils, and grease (FOG) bioindicators. Indicators of excess FOG include increase in the number of filamentous organisms that grow in the presence of excess FOG (a), presence of oil droplets under the india ink reverse stain, and an oily coating on floc particles (b).

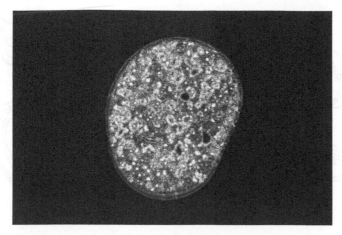

Figure 5.14   *Oil droplets. An oil droplet in the mixed liquor can be observed with an India ink reverse stain of mixed liquor. Because the india ink is an aqueous solution, the wastewater around the oil droplet stains black, while the oil droplet is clear.*

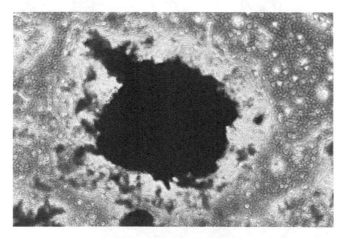

Figure 5.15   *Oil film on a floc particle. Oil is adsorbed to the surface of floc particles.*

smears of the mixed liquor. The larger the number of microscopic observations marked as "X," the greater the probability of the occurrence of that condition.

Second, if an unknown adverse operational condition should occur in an activated sludge process, all the "X"s in each row for "microscopic observation" that are identified during microscopic examinations of mixed liquor should be highlighted. The column(s) with the largest number of highlighted "X"s should be suspected as the probable adverse operational condition.

With increasing proficiency in microscopic techniques and identification of components of the mixed liquor biota food chain, the microscope can be used more efficiently as a diagnostic tool for process control and troubleshooting of the activated sludge process. Periodic use of microscopic examinations of mixed liquor, foam, and other wastewater samples can serve the following purposes:

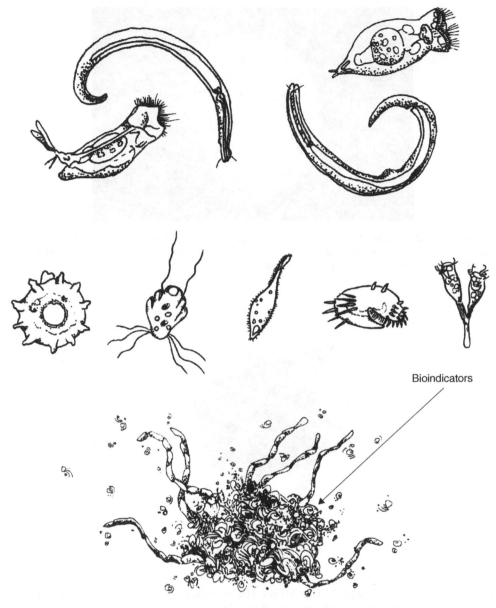

Bioindicators

Figure 5.16 *Zoogloeal growth bioindicators. Bioindicators of excess zoogloeal growth include the presence of easily observed amorphous zoogloeal growth or dendritic zoogloeal growth.*

TABLE 5.1 Adverse Operational Conditions and Microscopic Observations

Microscopic Observation	Adverse Operational Condition*												
	1	2	3	4	5	6	7	8	9	10	11	12	13
Sheared stalked ciliates							X						
Dispersed growth, increase	X			X	X	X	X	X	X		X		
Small floc, dominant							X		X				
Irregular floc, dominant							X						
Floating broken filaments							X		X				
Inactive protozoa	X								X				
Spherical floc, dominant									X				
Dispersed metazoa									X				
Weak floc under methylene blue	X			X	X	X		X	X				
Growth of low dissolved oxygen filaments	X												
Free-swimming stalked ciliates	X												
Regression in protozoan groups	X									X			
Growth of nutrient-deficient filaments			X										
India ink stain, positive			X										
Growth of low F/M filaments		X											
Growth of low pH filaments				X									
Growth of high pH filaments					X								
FOG covering floc												X	
Growth of FOG filaments												X	
Growth of septic/sulfide filaments						X							
Zoogloeal growth													X
Congealed and oval floc											X		
Floc, weak perimeter, and firm cores								X					
Stalked ciliates with bubbles										X			

*1—Low dissolved oxygen concentration; 2—low F/M; 3—low nutrients; 4—low pH; 5—high pH; 6—septicity/sulfides; 7—shearing action; 8—slug discharge of soluble cBOD; 9—surfactants; 10—toxicity; 11—heavy metal; 12—fats, oils, and grease (FOG); 13—zoogloeal growth.

- Compare healthy and unhealthy conditions
- Identify proper operational conditions
- Identify upset operational conditions
- Determine the success or failure of corrective measures
- Troubleshoot potential problems
- Identify causative factors for settleability problems
- Identify causative factors for loss of solids
- Identify effluent solids characteristics
- Identify causative factors for foam production
- Document impact of industrial discharges

6

The Microscope

Wastewater microscopy is the use of the compound microscope or the stereoscopic binocular microscope to identify the quality and quantity of the major living components or biota of the mixed liquor or other wastewater samples (Table 6.1). Typically, the compound microscope is used more often. The compound microscope may be bright-field or phase contrast.

The compound microscope used in the laboratory consists of two lens systems, the ocular lenses and the objective lenses (Figure 6.1). The compound microscope magnifies significant components of the biota that are too small to be seen with the unaided eye (Table 6.2), which can only see objects as small as 0.1 mm or 100 μm. Most bacteria are <2 μm.

Microscopic objects are measured in the metric system in micrometers or microns (μm) and nanometers (nm). A micron is a millionth of a meter (m) or a thousandth of a millimeter (mm). A nanometer is a thousandth of a micron. Bacteria, fungi, protozoa and rotifers typically are measured in microns. The bacterium *Escherichia coli* is approximately 2 μm × 1 μm in size. Viruses are measured in nanometers.

The lenses closest to the eye are the ocular lenses or eyepieces. Some microscopes have one ocular lens (monocular); some have two ocular lenses (binocular); and some have three ocular lenses (trinocular). Most compound microscopes are binoculars that help to reduce eyestrain. However, they do not provide stereoscopic vision. Trinocular microscopes allow a second person to observe a wastewater sample or a camera to be attached for photomicrography. Most ocular lenses magnify 10×. Some ocular lenses magnify 15×.

Most compound microscopes typically have three objective lenses. The lenses are located immediately above the "object" or specimen to be observed. The lenses

TABLE 6.1 Microscopic Biota of the Mixed Liquor

Actinomycetes (Nocardioforms)	Nematodes
Algae	Protozoa
Bacteria	Rotifers
Dispersed growth	Spirochetes
Filamentous organisms	Tetrads
Flatworms or gastrotriches	Water bears
Floc particles	Zoogloeal growth
Fungi	

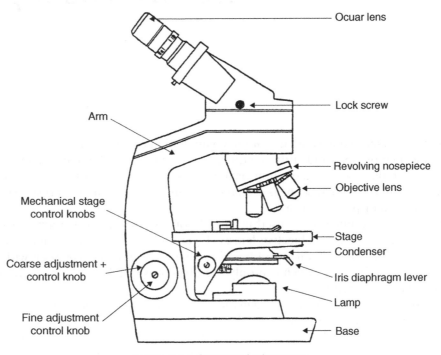

Figure 6.1 *Compound microscope.*

TABLE 6.2 Comparison of Bright-Field and Phase Contrast Microscopes

Feature	Microscope	
	Bright Field	Phase Contrast
Light source	Transmitted light	Transmitted light
Lenses	Optical lenses	Optical lenses and difraction plates
Image	Light objects on a bright background	Light waves in phase
Magnification, total	1000×	1000×
Cost	Less expensive	More expensive
Staining	Often required	Seldom required
Viewing of specimen	Alive or dead	Alive or dead

consist of the 10× (low power), 40× (high power), and 100× (oil immersion) lenses. The 100× or oil immersion lens is used by lowering or placing the lens in oil before focusing it on the specimen. Some microscopes have a 4× (scanning power) objective lens.

The objective lenses provide a "real" image of the specimen that is observed. The ocular lenses produce a "virtual" image of the specimen. Light rays from an illuminator or light source travel through a condenser that directs the rays through the specimen. The light rays then travel into the objective lens, where a "real" image of the specimen is formed on a mirror. The real image is magnified again by the ocular lenses. The ocular magnification produces an inverted or "virtual" image of the specimen, that is, a secondary mirror image in which left and right and top and bottom are reversed. Therefore, the virtual image of the specimen moves opposite of the direction in which a slide is moved.

The total power of magnification of the ocular and objective lens systems is obtained by multiplying the power of the ocular lens by the power of the objective lens that is used. For example, for the 40× or high-power objective lens, the total magnification would be 10 (ocular magnification) times 40 (high-power magnification) and would equal 400. The total power of magnification for each combination of ocular and objective lenses has general and specific uses for wastewater microscopy (Table 6.3).

With increasing total power of magnification the size of the specimen becomes larger, but the field of view becomes smaller (Figure 6.2). The field of view is the area that the microscopist is able to see with each power of magnification. Also, with increasing total power of magnification more light or increasing light intensity is required, and with deceasing total power of magnification less light or decreasing light intensity is required. If light intensity is not properly regulated, for example, too much light intensity at low-power magnification, translucent specimens may not be visible or specific features of a specimen may not be distinguished.

Two types of compound microscopes are commonly used in wastewater laboratories. These are the bright-field and phase contrast microscopes. In a bright-field microscope, light from a lamp is concentrated directly on the specimen or organism

TABLE 6.3 Total Powers of Magnification and Their General and Specific Uses in Wastewater Microscopy

Total Power of Magnification	Use
40×	Identify and observe entire body of large metazoa
100×	General overview of the bulk solution and biomass
	Identify protozoa to groups
	Perform protozoan and metazoan counts
	Identify small metazoa
	Determine location and relative abundance of filamentous organisms
	Identify interfloc bridging and open floc formation
400×	If needed, identify protozoa to groups
	If needed, identify protozoa to genus or species
	Detailed observations of floc particles
1000×	Identification of filamentous organism structure
	Identification of filamentous organism response to specific stains

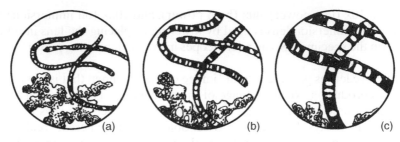

Figure 6.2   *Changing powers of magnification. When observing the same field of view at low-power magnification (a), high-power magnification (b), and oil immersion (c), the size of the image that one is viewing becomes larger with increasing power of magnification but the size of the area that one is viewing becomes smaller.*

to be observed. With this type of illumination, most organisms are difficult to observe because of their lack of contrast with the surrounding medium or wastewater. Therefore, careful regulation of light intensity and, often, the addition of an appropriate stain such as methylene blue must be used to better view the organism.

In a phase contrast microscope a special diaphragm in the condenser modifies a portion of the light that passes through the microscope. The light passes through the specimen at different speeds, that is, "out of phase." This modification exaggerates the refractive index of the specimen, making it possible to distinguish structural details that vary slightly in thickness or refractive properties. Phase contrast is based on the principle that specimens differ in refractive index from their surrounding medium and therefore bend some of the light waves that pass through them. This provides a clear image with increased degree of contrast over that obtained with bright-field microscopy. Phase contrast microscopy is used routinely for examining and identifying filamentous organisms that are responsible for settleability problems in the activated sludge process. Examples of cellular structures commonly examined with phase contrast microscopy include vacuoles and granules in protozoa and bacteria as well as flagella and cilia.

Phase contrast microscopy is available on most advanced microscopes. The microscopist is able to change to bright-field or phase contrast microscopy simply by rotating to a specific setting on the thumb wheel that is located immediately beneath the stage of the microscope (Figure 6.3). By matching the thumb wheel setting (10×, 40×, 100× or Ph 1, Ph 2, Ph 3) to the objective lens that is being used, the objective lens is used as a phase-contrast objective lens. The thumb wheel also has a "BF" or "0" setting that permits the use of each objective lens as a bright-field objective lens.

Although the ability to magnify a specimen many times is important when performing microscopic examinations of wastewater samples, so too is the ability to see clearly two specimens as separate objects. The latter ability is called the resolution or resolving power of the microscope.

The resolving power of the microscope is determined in part by the wavelength of light used for observing a specimen. Visible light has a wavelength of approximately 500 nm, while ultraviolet light has a wavelength of approximately 400 nm or less. The resolving power of the microscope increases as the wavelength of light

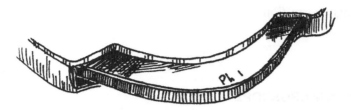

Figure 6.3 *Thumb wheel. Beneath the stage of the phase contrast microscope there is a thumb wheel with markings that typically are "0;" "10" or "Ph 1;" "40" or "Ph 2," and "100" or "Ph 3." By setting the thumb wheel to "0," any objective lens may be used as a bright-field lens. By matching the thumb wheel setting, for example, "10" or "Ph 1" and placing the low-power lenses or 10× lens in position, the lens can be used as a phase contrast lens.*

decreases. Therefore, ultraviolet light permits the detection of specimens not seen with visible light.

The resolving power of an objective lens refers to the size of the smallest specimen that can be observed with the lens. The resolving power is based on the wavelength of light used and the widest cone of light or numerical aperture (NA) that can enter the lens.

To observe a specimen clearly, sufficient light must enter the objective lens. This is not a problem for scanning-power, low-power, and high-power lenses. However, the oil immersion lens is very narrow, and most light does not enter the lens. Therefore, a drop of immersion oil is placed between the lens and the glass slide. Immersion oil has the same index of refraction as the glass slide. With the same index of refraction for immersion oil and the glass slide, light is maintained in a straight line as it passes through the glass slide, immersion oil, and objective lens.

Refraction or "distortion" of light occurs when light crosses a boundary between two media such as glass and air that have different refractive abilities. Refraction results in a change in the direction of the light. The degree of change in the light is the index of refraction or refractive index. If immersion oil was not used, the light would travel through the glass slide, air, and objective lens. This would result in much refraction of the light, that is, much of the light would not enter the objective lens, and the resolution of the specimens would decrease.

Best resolution is obtained when immersion oil having approximately the same index of refraction (1.6) as glass is used. Because oils do not evaporate when exposed to air for long periods of time, extended periods of time for microscopic examination are possible when using immersion oil.

FILTERS

Filters are the interchangeable circular colored glasses that are placed over the light source or lamp of the microscope. Filters are used to improve microscopic examinations of wastewater samples and enhance photomicroscopy. Applications of filters include

- Color effects
- Diffusion

- Mixed light correction
- Polarization
- Reflection

SELECTING A MICROSCOPE

The compound microscope is considered as standard equipment in many wastewater laboratories. However, there are many accessories and modifications that may be incorporated into the microscope for comfort, ease of handling, improved contrast or resolution, and photomicrography work. Therefore, careful consideration should be given to accessories and modifications when selecting a microscope. Several considerations would include the following:

- Use of binocular lenses to reduce eyestrain and fatigue
- Use of trinocular lenses to incorporate photomicrography ability
- Use of a 4× (scanning power) objective lens to examine and identify large microscopic metazoa and macroinvertebrates
- Use of mechanical stage control knobs for smooth and precise movement of microscope slides when performing scans or counts of protozoa and metazoa
- Use of filters for "cooling" the light and providing contrast in a field of view

Before purchasing a microscope, a microscope may be used on a trial basis. Many vendors of microscopes may permit an operator to use a microscope for a few weeks to determine whether the microscope is acceptable for purchase.

PARTS OF THE MICROSCOPE

Major parts of the microscope (Figure 6.1) and their functions include the following:

Ocular lens or eyepiece: usually 10× power (sometimes 15× power), magnifies an image 10 times.

Eye cups: rubber cups placed over the ocular lenses to block light from entering the sides of the ocular lenses when performing microscopic examinations.

Filters: circular colored glasses that are placed over the light source of the microscope to improve microscopic examinations and enhance photomicrography.

Thumb wheel or interpupillary adjustment: found in binocular microscopes, that is, microscopes having two ocular lenses; used to adjust the distance between the ocular lenses to match the interpupillary distance of the microscopist.

Lock screw: can be loosened or tightened to rotate and lock the position of the head of the microscope.

Head: holds the ocular lenses and tubes or barrels in which the lenses are inserted.

Arm: holds the head and stage; used along with the base to carry the microscope.

Revolving nosepiece: a rotating plate containing the objective lens; usually three or four objective lenses are mounted on the nosepiece. By rotating the nosepiece, a different objective lens can be placed in position over the specimen.

Objective lenses: magnify the specimen, and may consist of a scanning-power lens (4×), a low-power lens (10×), a high-power lens (40×), and an oil immersion lens (100×).

Slide holder or stage clip: consists of fixed and movable parts that secure the microscope slide upon the stage.

Mechanical stage: includes the slide holder and may be calibrated to relocate a specimen. The movement of the slide on the mechanical stage is controlled by the mechanical stage control knobs.

Mechanical stage controls or control knobs: usually consist of two knobs (large and small) that regulate the movement of the slide on the stage. The movement of the slide from left to right is regulated by one knob, while the other knob regulates the movement of the slide from top to bottom. Sometimes called the X and Y stage travel controls.

Stage: the flat area or table beneath the objective lenses that holds the microscope slide.

Stage aperture: allows light from the base of the microscope to reach the specimen.

Condenser: focuses light on the specimen and fills the objective lens with light.

Diaphragm lever: regulates the size of the iris opening in the stage aperture. By regulating the size of the iris opening, the diaphragm lever controls the amount of light entering the stage aperture.

Substage adjustment knob: raises and lowers the condenser, thereby adjusting the focusing of light and amount of light entering the objective lens.

Coarse adjustment knob: used to raise or lower the stage and rapidly bring the specimen into focus under scanning-power and low-power lenses.

Fine adjustment knob: used to raise or lower the stage and slowly bring the specimen into focus under high-power and oil immersion lenses.

Light or light bulb: illuminates the specimen.

Base: supports the microscope and is used along with the arm to carry the microscope.

FOCUSING

When not in use the microscope should be stored under a protective dust cover. Its electrical cord should be wrapped and secured around the base. The stage and condenser should be set at their lowest positions.

When carrying the microscope to and from the laboratory counter, both hands should be used. One hand should be placed beneath the base of the microscope, and one hand should be placed around the arm of the microscope.

Focusing with the microscope should be done according to the owner's manual. However, if the owner's manual is not available the following procedure is recommended for focusing and examining wastewater samples:

- Remove the dust cover, plug in the microscope, and turn on the power.
- Clean the ocular lenses and objective lenses with lens paper. Other papers may scratch the lens and leave fibers behind.
- Place a wet mount with coverslip on the stage and secure the wet mount with the stage clips or slide holder.
- Make sure the 10× objective lens is in position, that is, directly above the ring of light coming through the stage. If the 10× objective lens is not in position, rotate the nosepiece until the lens "clicks" into position. Place the objective lens in position by grasping and turning the nosepiece, not the objective lens.
- If the microscope has a substage condenser, set it to its uppermost position.
- Using the mechanical stage control knobs, move the wet mount until the edge or corner of the coverslip is at the middle of the light coming through the stage. This places the edge or corner of the coverslip within the field of view.
- While looking through the ocular lenses, adjust the thumb wheel for correct interpupillary distance.
- Adjust the focus by first looking into the right ocular lens with the right eye and focusing on the edge or corner of the coverslip or, better yet, a specimen close to the edge of the coverslip. Microscopy work may be performed without glasses.
- Next, look into the left ocular lens with the left eye and focus the image with the diopter ring.
- Raise the condenser to the bottom of the microscope slide and then slightly lower the condenser for maximum light.
- If needed, rotate the nosepiece until the 40× (high power) objective lens "clicks" into position above the wet mount.
- Adjust the aperture diaphragm with the diaphragm lever until there is sufficient light passing through the wet mount. If the microscope has a rheostat, use the lowest rheostat setting. The rheostat setting may need to be increased with increasing powers of magnification.
- The wet mount may now be scanned. Scanning is performed by moving the slide from side to side and from top to bottom, using the mechanical stage control knobs. Each field of view that is scanned should be examined by focusing through the field of view. Focusing through is done by slowly rotating the fine adjustment knob clockwise and counterclockwise. This raises and lowers the objective lens a short distance, permitting the microscopist to examine the top and bottom of the field of view. Note that objects will appear to be inverted (upside down and backwards). These inversions are caused by the optics of the microscope, and some practice will be required to compensate for these inversions.
- To view more detail, use the high-power (40×) lens, but always locate the specimen first by using the low-power (10×) lens. The specimen should be centered in the field of view and in focus before changing objective lenses. Most

microscopes have parafocal objectives, that is, if the specimen is in focus with the low-power lens, it should be nearly in focus with the high-power lens. Only a minor adjustment with the fine adjustment knob should be needed. Increase light intensity with increasing power of magnification and decrease light intensity with decreasing power of magnification.

- Remove the wet mount when finished with the microscopic examination.
- Clean the ocular and objective lenses with lens paper. Remove any oil.
- Rotate the 10× objective lens into position.
- Lower the stage.
- Turn off the power.
- Unplug and wrap the electrical cord around the base of the microscope.
- Place the dust cover over the microscope and place the microscope in its storage area or cabinet.

OIL IMMERSION

The use of the 100× objective lens or oil immersion lens requires the use of immersion oil. The tip of the 100× objective lens must be completely covered in the oil to reduce the distorting of light rays as they pass through the specimen to the objective lens. Less distortion of light rays occurs through immersion oil than through air. Focusing is done with the fine adjustment knob. The following procedure should be followed when using the 100× objective lens:

- With the 40× objective lens out of position, place a small drop of oil of immersion on the microscope slide above the area being examined. Then swing the oil immersion lens and 100× objective lens, into position.
- It may be necessary to increase the amount of light intensity again in order to properly view the specimen at 1000× total magnification.
- After viewing the specimen at 1000× total magnification, swing the immersion oil lens out of position.
- Clean the immersion oil with lens paper, and place the scanning-power (4×). or low-power (10×) objective lens into position.

7

Microscopic Measurements

Microscopic measurements, or accurately sizing microscopic specimens, is performed with a calibrated ocular micrometer. Specimens commonly measured during microscopic examinations of wastewater samples include algae, bacteria, dispersed growth, free-living nematodes, fungi, filamentous organisms, floc particles, and rotifers.

An ocular micrometer is a scale of lines (Figure 7.1) that may be etched in an ocular lens or in a disk (micrometer disk) inserted under an ocular lens. Although the length between the lines of the scale is uniform, the exact length under magnification is unknown. It must be calibrated for each total power of magnification, that is, 40× or scanning power, 100× or low power, 400× or high power, and 1000× or oil immersion. Calibration of the ocular micrometer is performed with a stage micrometer (Figure 7.1).

The stage micrometer is a slide with a scale usually containing large lines that are 0.1 mm apart and small lines that are 0.01 mm apart. These values will be labeled on the slide. The stage micrometer is placed in the microscope slide holder on the microscope stage. With the stage micrometer in position, each objective lens (4×, 10×, 40×, and 100×) is in turn placed over the stage micrometer. The scale of the ocular micrometer is calibrated at each power of magnification with the stage micrometer. Once the length of the distance between the lines on the ocular micrometer has been determined for each power of magnification, the stage micrometer is removed.

CALIBRATING THE OCULAR MICROMETER

To calibrate the ocular micrometer, the following procedure is recommended:

Microscopic Examination of the Activated Sludge Process, by Michael H. Gerardi
Copyright © 2008 John Wiley & Sons, Inc.

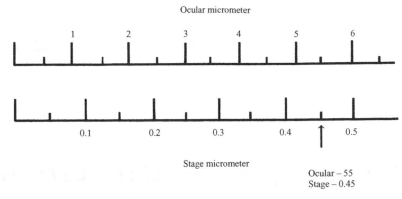

Figure 7.1 *Micrometers. To perform microscopic measurements, two micrometers are needed. The ocular micrometer is found or inserted in one tube of an ocular lens. The ocular micrometer remains in the tube. To calculate the distance in mm between the lines of the ocular micrometer, a stage micrometer is used. The stage micrometer is found on a stage micrometer microscope slide and is placed on the stage of the microscope. Starting with low-power magnification and then with each increasing power of magnification, the distance in mm between the ocular lines is calculated. However, the left end line or "zero" line of each micrometer must be aligned, and the micrometers must be superimposed by using the mechanical stage control knobs of the microscope. In this example, the 55 line of the ocular micrometer and the 0.45 line of the stage micrometer are the first lines to the left of each zero line that are aligned.*

(1) Place the scanning-power or 4× objective lens in position.

(2) Check to see whether the ocular lens of the microscope has an etched ocular micrometer or an inserted micrometer disk.

(3) If an etched ocular micrometer or micrometer disk already is in the ocular lens, rotate the scale until it is horizontal.

(4) If the microscope has no etched ocular micrometer or micrometer disk, obtain a micrometer disk and insert it under the ocular lens of the microscope. Rotate the scale until it is horizontal.

(5) Obtain a stage micrometer. Examine the stage micrometer against the laboratory light, and note the tiny scale etched onto it. Place the stage micrometer in the microscope slide holder on the microscope stage. Center the scale over the stage aperture directly beneath the 4× objective lens and illuminate the scale with reduced light.

(6) With the coarse adjustment knob, minimize the distance from the stage micrometer to the 4× objective lens. Bring the stage micrometer into focus and center it in the field of view. Adjust the microscope so that optimum light is obtained, so both ocular and stage micrometers are sharp and clear.

(7) Turn the ocular lens until the lines of the ocular micrometer are parallel with those of the stage micrometer.

(8) On your left side of the field of view, align the two scales, so that the first line of the stage micrometer is centered and directly over the first line of the ocular micrometer. The "zero" points of both scales must coincide and must be parallel.

TABLE 7.1   Microscope Calibration

Objective	No. of Stage Spaces			No. of Ocular Spaces			mm or µm Value of One Ocular Space			Average value of one ocular space
	1	2	3	1	2	3	1	2	3	
4×										
10×										
40×										
100×										

(9)   From the first line of each scale that has been centered, carefully scan the scale of the stage micrometer. Scan it until another line of the stage micrometer scale is centered and directly above another line of the scale of the ocular micrometer.

(10)   Note and record the number of lines that coincide exactly, that is, how many spaces (divisions) are there between these matching lines?

Number of stage spaces: _____

Number of ocular spaces: _____

(11)   Divide the number of stage spaces by the number of ocular spaces.

(12)   Since each stage space equals 0.01 mm, each ocular space equals 0.01 mm times the mathematical result from step 11 or _____ mm. This is the distance in mm for each space of the ocular micrometer at 40× total power magnification. Because 1 mm equals 1000 µm, the number of mm obtained may be multiplied by 1000 to obtain the distance in µm for each space.

(13)   Using this procedure, calibrate in order the ocular micrometer with low-power, high-power, and oil immersion objective lenses. To lessen the error, make three different readings with each objective lens and take the average (Table 7.1).

Once the ocular micrometer has been calibrated for each objective lens, the space between each line of the ocular micrometer may be recorded on an index card and the card may be taped to the arm of the microscope. The index card serves as a reference to permit any microscopist to measure a specimen with any objective lens. The obtained calibrations are accurate only when the same microscope, same ocular lens, and same objective lenses are used.

OLYMPUS® WASTEWATER RETICLE

The Olympus® wastewater reticle (OWR™) is a customized "bull's-eye" pattern ocular micrometer designed specifically for measuring floc particles. The reticle does not require calibration and permits the operator to quickly estimate the size of floc particles when using the 10× phase-contrast objective lens.

8

The Stereoscopic Binocular Microscope

The stereoscopic binocular microscope (Figure 8.1) or stereomicroscope provides low-power magnification for improved examination of specimens such as blood-worms and tubifex that are too large for examination with the compound microscope and too small for detailed observations with the unaided eye. Unlike the compound microscope, the image of the specimen seen through the stereomicroscope is not inverted. The stereomicroscope also provides for three-dimensional viewing of a specimen. The stereoscopic binocular microscope is known also as the dissecting microscope, because it is used in biology laboratories for the dissection of specimens such as earthworms and grasshoppers.

The stereomicroscope consists of two lens systems—the ocular and the objective. There are two ocular lenses (binocular), and each lens focuses on the same field of view at a different angle. This permits stereoscopic or three-dimensional viewing of a specimen.

Most stereomicroscopes have 10× ocular lenses that can be adjusted for interpupillary distances. The objective lenses usually range in power from 3× to 7×. The total power of magnification is the product of the power of magnification of the ocular lenses times the power of magnification of the objective lens, for example, 10× ocular lenses multiplied by a 7× objective lens equals 70× total magnification.

The stereomicroscope has either an opaque or a transparent glass stage plate where the specimen is placed. When equipped with a transparent stage plate and a substage, the stereomicroscope can be used to view objects in transmitted as well as reflected light.

Transmitted light is used to view translucent specimens. Reflected light is used to view opaque specimens. Together, transmitted light and reflected light can be used to develop contrasting conditions on a variety of specimens. Reflected light is better for color and depth perception than transmitted light.

Microscopic Examination of the Activated Sludge Process, by Michael H. Gerardi
Copyright © 2008 John Wiley & Sons, Inc.

Figure 8.1 Stereoscopic binocular microscope.

Focusing is achieved by turning the focusing knob that raises and lowers the body tube. Most stereomicroscopes have variable or zoom magnification. Increasing and decreasing magnification is performed by rotating a knob on the top of the stereomicroscope or rotating a ring around the center of the body. The specimen remains in focus while the power of magnification changes.

USING THE STEREOSCOPIC BINOCULAR MICROSCOPE

Focusing with the stereoscopic binocular microscope should be done according to the owner's manual. However, if the owner's manual is not available the following procedure is recommended for focusing and examining specimens:

- Place the stereomicroscope on a hard, stable table or laboratory counter.
- Plug the stereomicroscope power cord into a suitable grounded electrical outlet.
- Select the stage plate that provides the greatest contrast between the specimen and the background.
- Place the specimen on the stage plate directly below the center of the body of the stereomicroscope.
- Turn on the appropriate light source and set the stereomicroscope at the lowest power of magnification.
- While looking through the ocular lenses with both eyes, adjust the distance between the ocular lenses until a complete circle of light is obtained.
- While looking through the ocular lenses, bring the specimen into sharp focus by turning the focusing knob.
- Focus adjustable ocular lenses to compensate for eye vision differences. With one eye closed, look through the nonadjustable ocular lens and focus sharply

on the specimen with the focusing knob. Then look through the adjustable ocular lens with the other eye and rotate the adjustable ocular lens's milled cuff to obtain a clear focus.

CARE OF THE STEREOSCOPIC BINOCULAR MICROSCOPE

The following instructions are provided for proper care and maintenance of the stereoscopic microscope:

- Remove dust and debris from the surface of lenses with an ear syringe or brush the lenses with a clean camel's hair brush.
- Clean the lenses with lens paper.
- When finished using the stereomicroscope, return the body to the lowest position, turn off the lights, unplug the electrical cord, wrap and secure the cord around the arm of the stereoscope, and place the dust cover over the stereoscope.

9

Equipment and Supplies

In addition to a bright-field microscope or a phase contrast microscope, appropriate stains, and immobilizing agents, the following required and recommended items should be available for microscopic examination of wastewater samples:

- Ajax® or similar compounds for cleaning oily film off of new slides
- Antibacterial hand lotion
- Antibacterial soap
- Beakers, 250 mL
- Beaker with disinfectant for disposal of used slides and coverslips
- Bibulous paper for blotting and drying slides
- Cleaning agent for ocular lenses and objective lenses. Numerous cleaning agents are available, including distilled water, acetone, ethanol, 90% isopropyl alcohol (not rubbing alcohol), and xylene.
- Clothespin for holding slides during staining procedures
- Coverslips, 22 × 22 mm (standard size for performing protozoa and metazoa counts). A no. 1½ coverslip is recommended.
- Disposable plastic droppers, pipettes, or straws for mixing and dispensing wastewater samples onto slides
- Dropper bottles for holding and dispensing prepared stains
- Filters for cooling light intensity or providing contrast
- Immersion oil
- Immobilizing agent [Detain™, Protosol™, mercuric chloride ($HgCl_2$), and nickel sulfate ($NiSO_4$)] for protozoa

- Kimwipes®
- Lens cleaning kit
- Lens paper
- Microscope slides, 25 × 75 mm
- Ocular micrometer
- Parafilm™ and scissors to seal dropper bottles of stains
- Photomicrography equipment or drawing paper, pencil, eraser, and colored pencils
- Pictorial keys for protozoa, metazoa, and macroinvertebrates
- Protective gloves, disposable
- Q-tips® for applying cleaning agent
- Refrigerator, 4 °C
- Rinse or wash bottle of distilled water for rinsing stains
- Sample bottles, wide mouth and plastic with lids
- Stage micrometer
- Taxonomic key for filamentous organisms
- Toothpicks for mixing stains with wastewater samples on slides
- Wax pencil for labeling slides
- Worksheets and rating tables

10

Wet Mounts and Smears

Two slide preparations are used for microscopic examinations of wastewater samples. These preparations are the wet mount and the smear. The wet mount is used more often than the smear.

To prepare a wet mount (Figure 10.1), the following procedure is recommended:

(1) Place a clean paper towel on the work area of the laboratory counter.

(2) Clean the surface of a 25×75 mm microscope slide and place the slide on the paper towel.

(3) Shake a sealed container of mixed liquor or wastewater and transfer a portion of the mixed liquor to a clean beaker. Stir the mixed liquor in the beaker. If protozoan work is to be performed, aerate the mixed liquor with a pipette and a pipetting bulb before transferring the mixed liquor from the container to the beaker and from the beaker to the microscope slide.

(4) With an eyedropper, pipette, or straw, place a drop of mixed liquor on the center of the microscope slide.

(5) Clean the surface of a 22×22 mm coverslip.

(6) Place the coverslip on the drop of mixed liquor in the following manner:

    a. Without touching the drop of mixed liquor, hold the coverslip at a 45° angle on the microscope slide between the thumb and the forefinger of the dominant hand. The 45° angle faces the drop of mixed liquor.

    b. Slowly slide the coverslip toward the drop of mixed liquor and allow the mixed liquor to touch and spread along the edge of the coverslip.

    c. Release the coverslip and allow it to fall on the mixed liquor. No air bubbles should be trapped under the coverslip.

Microscopic Examination of the Activated Sludge Process, by Michael H. Gerardi
Copyright © 2008 John Wiley & Sons, Inc.

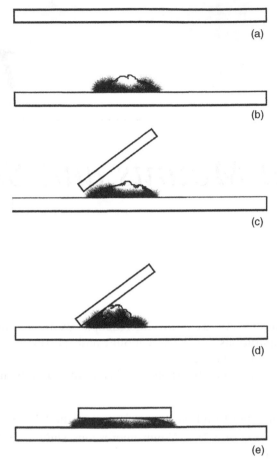

Figure 10.1 *Wet mount of mixed liquor. To prepare a wet mount of mixed liquor, a clean microscope slide is placed on a paper towel on the laboratory counter* (a). *A drop of mixed liquor is placed on the center of the microscope slide* (b). *A coverslip is placed on the microscope slide at a 45° angle* (c). *The coverslip is drawn toward the mixed liquor* (d), *and when the coverslip touches the mixed liquor, the coverslip is dropped over the mixed liquor* (e).

    d. Place a clean sheet of tissue over the coverslip. With a blunt object, gently press down on the coverslip to remove excess mixed liquor.

    e. Remove the tissue and discard it appropriately. This leaves approximately 0.05 mL of mixed liquor under the coverslip.

    f. Use a wax pencil to label the slide on the left side with the date and sample.

A smear of mixed liquor or foam is performed to permit staining of the smear's biological components, their structural features, and their reactions (negative or positive) to specific stains such as the Gram stain and Neisser stain. Mixed liquor and foam smears are prepared differently.

To prepare a smear of mixed liquor (Figure 10.2), the following procedure is recommended:

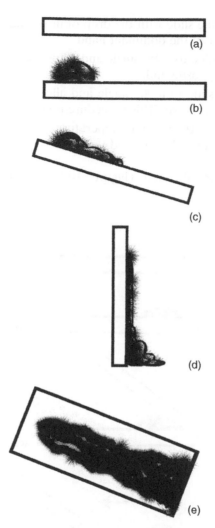

Figure 10.2 *Smear of mixed liquor. To prepare a smear of mixed liquor, a clean microscope slide is placed on a paper towel on the laboratory counter (a). A drop of mixed liquor is placed on one end of the slide (b). The end of the slide with the mixed liquor is slowly lifted to a 45° angle and the mixed liquor is allowed to run down the slide (c). When the mixed liquor reaches the opposite end of the slide, the slide is lifted to a 90° angle and the excess mixed liquor is permitted to flow off the slide onto the paper towel (d). The smear is allowed to air dry. This procedure produces a cone-shaped thin smear of mixed liquor (e).*

(1) Place a clean paper towel on the work area of the laboratory counter.
(2) Clean the surface of a 25×75 mm microscope slide and place the slide on the paper towel.
(3) Shake a sealed container of mixed liquor or wastewater and transfer a portion of the mixed liquor to a clean beaker. Stir the mixed liquor in the beaker.
(4) With an eyedropper, pipette, or straw, place a drop of mixed liquor on one end of the microscope slide.

(5) Hold the end of the slide with the drop of mixed liquor between the thumb and the forefinger of the dominant hand.

(6) Slowly raise the slide to a 45° angle and allow the mixed liquor to run down the slide to the opposite end.

(7) Slowly raise the slide to a 90° angle and allow any excess mixed liquor to flow onto the paper towel. This procedure produces a cone-shaped smear.

(8) Allow the slide to dry at room temperature. After the slide has dried completely, it can be stained as desired for microscopic examination

To prepare a smear of foam (Figure 10.3), the following procedure is recommended:

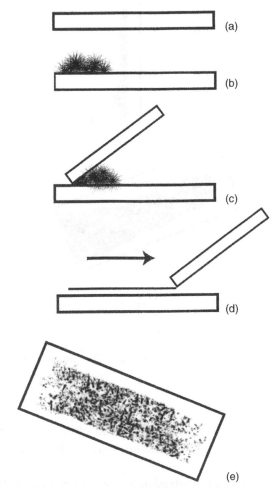

(a)

(b)

(c)

(d)

(e)

Figure 10.3   *Smear of foam. To prepare a smear of foam, a clean microscope slide is placed on a paper towel on the laboratory counter (a). A sample of foam is placed on one end of the slide (b). Another clean microscope slide is placed on the foam-laded slide at a 45° angle (c) and pulled across the foam to the opposite end of the foam-laden slide (d). The smear is allowed to air dry. This procedure produces an thin smear of foam across the surface of the slide (e).*

(1) Place a clean paper towel on the work area of the laboratory counter.

(2) Clean the surface of a 25 × 75 mm microscope slide and place the slide on the paper towel.

(3) Transfer a small quantity of foam on the tip of a stirring rod or straw to one end of the microscope slide.

(4) Hold the slide on the paper towel with the thumb and forefinger of the non-dominant hand at the end of the slide that has the sample of foam.

(5) With the thumb and the forefinger of the dominant hand, hold another clean microscope slide at a 45° angle near the end of the microscope slide that has the sample of foam.

(6) While securely holding both slides, draw the 45° angle slide toward the opposite end of the microscope slide. This produces a thin, wide smear of foam.

(7) Allow the slide to dry at room temperature. After the slide has dried completely, it can be stained as desired for microscopic examination.

SCANNING A WET MOUNT

To scan a wet mount, the following procedure is recommended:

(1) Place the wet mount on the stage of the microscope and secure the wet mount in position with the slide holder or stage clips.

(2) With the mechanical stage control knobs, position the wet mount so that you can view a corner of the coverslip, using the 10× objective lens.

(3) Adjust the light intensity and focus on the corner of the coverslip.

(4) Examine the first field of view at the corner of the coverslip. Be sure to focus through on this field of view and all fields of view. Focusing through consists of slowly turning the fine adjustment knob clockwise and counterclockwise. This raises and lowers the objective lens only a small distance and permits an examination of the components of the wet mount at the top, middle, and bottom levels of the wet mount.

(5) After examining the first field of view, move the slide laterally with the mechanical stage control knob until the second field of view is in position. Be sure that the edges of the first and second fields of view overlap.

(6) Continue moving the slide laterally until a row of fields of view have been examined. Then, using the mechanical stage control knob, move the slide up or down to the next row of fields of view and scan this row. Be sure that the top and bottom edges of each row of fields of view overlap.

(7) Continue the scan until the entire wet mount has been scanned.

FIELD OF VIEW

A field of view is the circular area that is observed through the microscope. With increasing power of magnification, the area of the field of view becomes smaller. Also with increasing power of magnification, more light intensity is required for proper examination of the field of view.

11

Staining Techniques

There are seven staining techniques that are used in performing microscopic analyses of mixed liquor (Table 11.1). These techniques include:

- Gram stain
- India ink reverse stain
- Methylene blue stain
- Neisser stain
- PHB stain
- Safranin stain
- Sheath stain

The staining techniques are used to provide contrast between specific components of the mixed liquor and the surrounding field of view, evaluate specific conditions in floc particles, and identify filamentous organisms to name or type number. In addition to these staining techniques the sulfur oxidation test or "S" test can be performed to assist in the identification of filamentous organisms.

GRAM STAIN

The Gram stain is a differential stain that separates bacteria into two groups based upon the response of the bacterial cell wall to a series of chemical solutions. These groups are Gram-negative (red) and Gram-positive (blue) bacteria. The Gram stain is used to identify filamentous organisms to name or type number. There are several

TABLE 11.1   Staining Techniques for the Examination of Mixed Liquor

Technique	Component Examined	Slide Preparation	Microscope Requirement	Total Power of Magnification
Gram	Filamentous organisms	Smear	Bright-field	1000×
India ink	Floc particles	Wet mount	Phase contrast	100× or 1000×
Methylene blue	All components	Wet mount	Bright-field or phase contrast	100×
Neisser	Filamentous organisms	Smear	Bright-field	1000×
PHB	Filamentous organisms	Smear	Bright-field	1000×
Safranin	Floc particles	Smear	Bright-field	100×
Sheath	Filamentous organisms	Wet mount	Phase contrast	1000×

Gram staining procedures. Perhaps the procedure that is most often used is the modified Hucker method.

Procedure:
Gram Stain, Modified Hucker Method

Solutions:
Prepare or purchase (Gram stain kit) the following four solutions:

Solution 1
Prepare A and B separately and then mix:

A

Crystal violet	2 g
Ethanol, 95%	20 mL

B

Ammonium oxalate	0.8 g
Distilled water	80 mL

Solution 2

Iodine	1 g
Potassium iodide	2 g
Distilled water	300 mL

Solution 3

95% Ethanol

Solution 4

Safranin O (2.5% in 95% ethanol)	10 mL
Distilled water	100 mL

Steps

1. Prepare a thin smear of mixed liquor on a microscope slide and permit the smear to air dry completely.
2. Stain the smear for 1 minute with Solution 1 and then rinse the slide with distilled water.
3. Stain the smear for 1 minute with Solution 2 and then rinse the slide with distilled water.

4. Holding the slide at a 45° angle, decolorize the slide with Solution 3 (95% ethanol) by adding the ethanol drop by drop to the smear for 30 seconds. Take care not to over-decolorize. Blot the slide dry.

5. Stain the smear for 1 minute with Solution 4 and then rinse the slide with distilled water and blot dry.

6. Examine the stained smear under oil immersion (1000× total magnification), using bright-field microscopy. Blue filamentous organisms are Gram-positive, while red filamentous organisms are Gram-negative.

INDIA INK REVERSE STAIN

An india ink reverse stain is used to determine the probability of a nutrient deficiency in the activated sludge process. With phase contrast microscopy the stain reveals the relative amount of stored food or insoluble polysaccharides that are deposited in the floc particles. The larger the quantity of polysaccharides that are present, the greater the probability of a nutrient deficiency in the activated sludge process.

India ink or nigrosine is an aqueous suspension of carbon black particles. When a drop or two of India ink is mixed with a small drop of mixed liquor on a microscope slide, the carbon black particles in the ink turn the bulk solution black and quickly penetrate into floc particles. As the carbon black particles penetrate from the perimeter of the floc particles to the core of the floc particles, bacterial cells become black or golden-brown in color. Stored food or polysaccharides block the penetration of the carbon black particles, and the stored food appears white under phase contrast microscopy. The larger the white area in the floc particles is, the greater the probability is that a nutrient deficiency in the bulk solution or activated sludge process exists.

India ink reverse stains that are reflective of a nutrient deficiency, that is, the area of most floc particles is white, are referred to as "positive." India ink reverse stains that are reflective of a nutrient adequate condition, that is, the area of most floc particles is black or golden-brown, are referred to as "negative." However, many "negative" floc particles may contain some small areas of white or stored food and produce a "spotty" white appearance.

There are two conditions that may produce "false" positives for the India ink reverse stain. These conditions are the encapsulation of bacteria as a result of toxicity and the undesired growth of zoogloeal organisms or viscous floc.

Some operators may find it difficult to locate floc particles when they are stained with India ink. To more easily find floc particles, the operator should focus on a corner or edge of the coverslip and then slowly scan the wet mount of mixed liquor and India ink. As white areas appear in a field of view during the scan the operator should focus-through the field of view to bring the floc particles in and out of focus. This allows the operator to see the perimeter of the floc particles and the relative amount of area of the floc particles that is white.

Procedure:
India Ink Reverse Stain

Solution:
India ink (aqueous solution of carbon black particles) or nigrosine

Steps

1. Mix one or two drops of India ink and one drop of mixed liquor on a microscope slide.

2. Place a coverslip on the India ink-mixed liquor sample and observe the sample at 1000× (oil immersion) under phase contrast microscopy.

3. Be sure that the floc particles that are being examined are surrounded by a black field of view.

4. In "nutrient-adequate" mixed liquor, the carbon black particles penetrate the floc particles almost completely, at most leaving only a few "spots" of white. This is a negative reaction to the India ink reverse stain.

5. In "nutrient-deficient" mixed liquor, large amounts of polysaccharides (produced through a nutrient deficiency) are present. The polysaccharides prevent the penetration of the carbon black particles. This results in the appearance of large white areas in the floc particles. This is a positive reaction to the India ink reverse stain.

METHYLENE BLUE

Methylene blue is used to provide contrast between organisms and their surrounding environment and to identify specific structural components in organisms such as the flagellum and the contractile filament in protozoa. Methylene blue enables an operator to more easily observe filamentous organisms, floc particles, metazoa, protozoa, and zoogloeal growth and to evaluate the strength of floc particles.

Procedure:
Methylene Blue Stain

Solution:
Prepare the following solution:

Methylene blue	0.01 g
Absolute alcohol	100 mL

Steps

1. Using dilute quantities of the stain, add a drop at the edge of the coverslip of a wet mount of mixed liquor and allow it to seep under the coverslip. Or, add a drop of methylene blue to the mixed liquor on a microscope slide, mix with a toothpick, and then add a coverslip. Do not overstain.

2. Examine the wet mount with either bright-field or phase contrast microscopy.

NEISSER STAIN

Like the Gram stain, the Neisser stain is a differential stain that separates bacteria into two groups based on their response to two stains. These groups are Neisser-negative (light brown to yellow) and Neisser-positive (gray-blue). The Neisser stain is used to identify filamentous organisms to name or type number.

Procedure:
Neisser Stain

Solutions:
Prepare the following two solutions:

Solution 1
Prepare A and B separately, then mix A and B at 2:1

A

Methylene blue	0.1 g
Acetic acid	5 mL
Ethanol (95%)	5 mL

B

Crystal violet 10% (in 95% ethanol)	3.3 g
Ethanol (95%)	6.7 g
Distilled water	100 mL

Solution 2

Bismarck brown or chrysoidine Y, 1% aqueous	33.3 mL
Distilled water	66.7 mL

Steps

1. Prepare a thin smear of mixed liquor on a microscope slide and permit the smear to air dry completely.
2. Stain the smear for 15 seconds with fresh (refrigerated and less than 6 months old) Solution 1 consisting of two parts of A and one part of B. Subsequently, allow the excess solution to flow from the slide.
3. Stain the smear for 45 seconds with Solution 2.
4. Rinse the slide with distilled water and allow the water to flow down the back of the slide. Blot the slide dry.
5. Examine the stained smear under oil immersion (1000×) with bright-field microscopy. Light brown to yellowish filamentous organisms are Neisser-negative, while gray-blue filamentous organisms are Neisser-positive.

PHB STAIN

Some filamentous organisms store food as intracellular "starch" granules. The granules are properly known as poly-β-hydroxybutyrate (PHB). Because the presence or absence of these granules can be detected by the PHB staining technique, the PHB stain can be used as a differential stain to help identify filamentous organisms to name or type number (Table 11.2).

Procedure:
PHB (β-polyhydroxybutyrate) Stain

Solutions:
Prepare the following two solutions:

Solution 1:

Sudan Black B (IV)	0.3% w/v in 60% ethanol

Solution 2:

Safranin O	0.5% w/v aqueous

TABLE 11.2   Filamentous Organisms that have PHB Granules

Beggiatoa spp.
Microthrix parvicella
Nocardioforms
Nostocoida limicola
Sphaerotilus natans
Type 1701
Type 021N

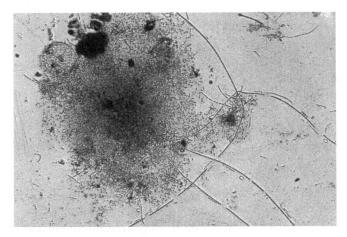

Figure 11.1   *Safranin stain. The Safranin stain may be used to detect the rapid young growth of bacterial cells on the perimeter of floc particles as a result of a slug discharge of soluble cBOD. Under the Safranin stain of a smear of mixed liquor the young bacterial cells are found loosely flocculated on the perimeter (light red), while the old bacterial cells are found tightly flocculated in the core (dark red).*

Steps

1. Prepare a thin smear of mixed liquor on a microscope slide and permit the smear to air dry completely.
2. Stain the smear for 10 minutes with Solution 1, adding more solution if the slide starts to dry, and then rinse with distilled water.
3. Stain the smear for 10 seconds with Solution 2 and then thoroughly rinse the slide with distilled water. Blot the slide dry.
4. Examine the stained smear under oil immersion (1000×) with bright-field microscopy. PHB granules appear as intracellular blue-black granules, while the cytoplasm appears pink or clear.

SAFRANIN STAIN

The safranin stain may be used to determine whether bacterial cells in a floc particle are tightly or loosely aggregated (Figure 11.1). For example, old cells secrete a relatively small quantity of polysaccharides during growth. Therefore, old cells are tightly aggregated. Under the safranin stain the growth of old bacterial cells is dark red, with little distance between the bacterial cells. Young cells secrete a copious quantity of polysaccharides during growth. Therefore, young cells are loosely aggregated. Under the safranin stain the growth of young cells is light red or pink, with much distance between the bacterial cells. The rapid young growth of bacterial cells from a slug discharge of soluble cBOD can be seen in floc particles under a safranin stain. The core of the floc particle is red with tightly aggregated cells, while the perimeter of the floc particles is light red or pink with loosely aggregated cells.

Procedure:
Safranin Stain

Solution:
Use Solution 4 from the Gram stain or prepare the following:

Safranin O (2.5% in 95% ethanol)	10 mL
Distilled water	100 mL

Steps

1. Prepare a thin smear of mixed liquor on a microscope slide and permit the smear to air dry completely.
2. Stain the smear for 1 minute with safranin O and then rinse the slide with distilled water and blot dry.
3. Examine the stained smear under oil immersion (1000×) with bright-field microscopy. Old bacterial growth is red and tightly aggregated, while young growth is light red or pink and loosely aggregated.

SHEATH STAIN

Some filamentous organisms possess a sheath or transparent protective covering that runs the entire length of the filamentous organism. Because the presence or absence of a sheath can be detected by the sheath staining technique, the sheath stain can be used as a differential stain to help identify filamentous organisms to name or type number (Table 11.3).

Procedure:
Sheath Stain

Solution:
Crystal violet
0.1% w/v aqueous

TABLE 11.3   Sheathed Filamentous Organisms

Haliscomenobacter hydrossis
Sphaerotilus natans
Thiothrix spp.
Type 0041
Type 0675
Type 1701
Type 1851

Steps

1. Mix one drop of mixed liquor and one drop of crystal violet solution with a toothpick on a microscope slide.
2. Place a coverslip on the mixed liquor-crystal violet sample and observe under oil immersion (1000×) with phase contrast microscopy. Filamentous cells stain deep violet, while the sheath stains pink or remains clear.

SULFUR OXIDATION TEST OR "S" TEST

Testing filamentous organisms for the ability or inability (presence or absence of specific enzymes) to use substrates (cBOD) or oxidize inorganic compounds can be used to identify filamentous organisms to name or type number. A commonly used biochemical reaction or test for the identification of filamentous organisms in mixed liquor is the sulfur oxidation test or "S" test.

The sulfur oxidation test is performed on a mixed liquor sample to determine whether filamentous organisms within the mixed liquor are capable of oxidizing sulfur and storing it as granules within the cytoplasm of individual cells. Some filamentous organisms such as *Beggiatoa* normally contain sulfur granules within their cytoplasm in vivo, that is, in their normal environment without applying the sulfur oxidation test. But some filamentous organisms produce detectable sulfur granules only after applying the sulfur oxidation test in vitro. Type 0092 and type 021N are examples that produce sulfur granules under the "S" test.

To determine whether filamentous organisms within the mixed liquor are capable of oxidizing sulfur, sulfur granules must be detected in vivo or in vitro, after applying the sulfur oxidation test.

Procedure:
Sulfur Oxidation Test

Solution:
Na_2S Solution
200 mg Na_2S per 100 mL of distilled water

Steps

1. Mix 15 mL of mixed liquor with 15 mL of Na_2S solution in a beaker. Allow the treated mixed liquor to stand for 15 minutes. Stir periodically to keep the solids in suspension.
2. After the 15-minute standing period, examine the filamentous organisms at 400× total power magnification under phase contrast microscopy. Determine whether sulfur granules were stored in the cytoplasm of the filamentous organisms. Sulfur granules are highly refractive and can be easily observed under phase contrast microscopy.

Part III

The Bulk Solution

12

Dispersed Growth

For the purposes of this text, dispersed growth is considered to be spherical floc particles that are ≤10 μm in diameter that are observed at 100× total magnification. Dispersed growth may be observed in a wet mount of mixed liquor under phase contrast microscopy (Figure 12.1) or a wet mount of mixed liquor with a drop of methylene blue under bright-field microscopy (Figure 12.2). Dispersed growth is associated with numerous operational conditions (Table 12.1).

The quantity of dispersed growth in the bulk solution may be rated as "insignificant," "significant," or "excessive." The evaluation of dispersed growth consists of

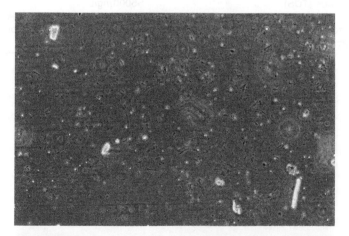

Figure 12.1 *Dispersed growth under phase-contrast microscopy.*

Figure 12.2 Dispersed growth under methylene blue and bright-field microscopy.

TABLE 12.1 Operational Conditions Associated with the Presence of Easily Observable Dispersed Growth

Operational Condition	Description or Example
Cell bursting agent	Lauryl sulfate
Colloidal floc	Nondegrading or slowly degrading colloids
Elevated temperature	>32 °C
Foam production	Foam-producing filamentous organisms
Increase in percent MLVSS	Accumulation of fats, oils, and grease
Lack of ciliated protozoa	<100/mL
Low dissolved oxygen concentration	<1 mg/L for 10 consecutive hours
Low pH/high pH	<6.5/>8.5
Nutrient deficiency	Usually nitrogen or phosphorus
Salinity	Excess sodium and/or potassium
Septicity	ORP < −100 mV
Shearing action (excess turbulence)	Surface aeration
Slug discharge of soluble cBOD	3× typical soluble cBOD
Surfactant	Anionic detergents
Total dissolved solids (TDS)	>5000 mg/L
Toxicity	Return activated sludge (RAS) chlorination
Viscous floc or zoogloeal growth	Rapid proliferation of floc-forming bacteria
Young sludge age	>3 days MCRT

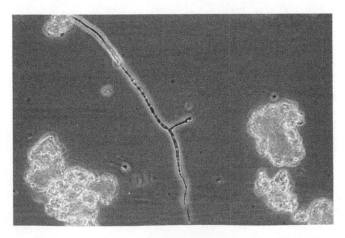

Figure 12.3 Dispersed growth at an "insignificant" rating.

Figure 12.4   *Dispersed growth at a "significant" rating.*

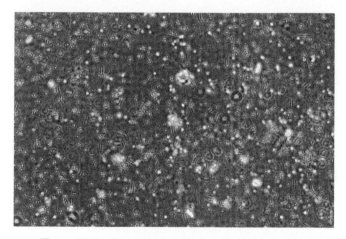

Figure 12.5   *Dispersed growth at an "excessive" rating.*

scanning several fields of view at 100× total magnification and evaluating the relative
abundance of dispersed growth according to the following ratings:

"Insignificant" (Figure 12.3)
Little dispersed growth
"Less than 20 particles per field of view"

"Significant" (Figure 12.4)
Much dispersed growth
"Tens of particles per field of field," for example, 10, 20, 30 …

"Excessive" (Figure 12.5)
Copious dispersed growth
"Hundreds of particles per field of view," for example, 100, 200, 30 …

13

Particulate Material

Particulate material can be found in a large variety of shapes, sizes, and colors in the activated sludge process. Particulate material consists of inert (nonliving) wastes that are either slowly biodegradable or nonbiodegradable. An example of biodegradable particulate material is plant fibers or cellulose (Figure 13.1). Examples of nonbiodegradable particulate material are plastic resins (Figure 13.2) and granular activated carbon (Figure 13.3).

In a healthy activated sludge process most particulate material is found within floc particles or extending into the bulk solution from the perimeter of floc particles. Particulate material is adsorbed to floc particles by compatible charge between the particulate material and the floc particle or through the coating action of secretions from higher life-forms, especially ciliated protozoa. These secretions coat and remove not only particulate material but also colloids from the bulk solution and place them on the floc particle. Proteins found in fecal waste, dairy wastewater, and slaughterhouse wastewater are examples of colloids.

In an unhealthy activated sludge process there may be an increase in the relative abundance of particulate material in the bulk solution. This increase usually is associated with an adverse operational condition that is responsible for the interruption of floc formation (Table 13.1). Interruption of floc formation results in the release of particulate material from floc particles or the failure of the floc particles to adsorb particulate material.

The relative abundance of particulate material in the activated sludge process may be rated as "insignificant" or "significant." The rating for a healthy activated sludge process is "insignificant," while the rating for an "unhealthy" activated sludge process is "significant." The evaluation of particulate material consists of scanning

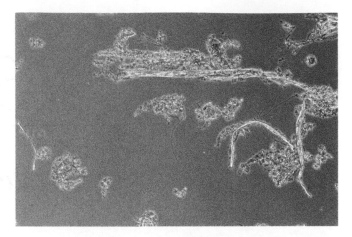

Figure 13.1 Fibrous particulate material.

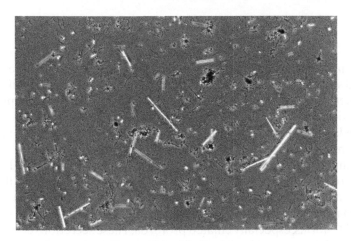

Figure 13.2 Plastic resins.

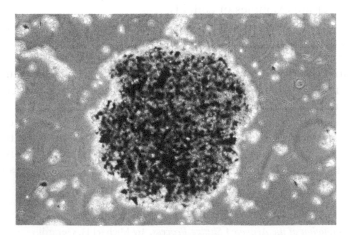

Figure 13.3 Granular activated carbon in a floc particle.

TABLE 13.1   Operational Conditions Associated with the Presence of Easily Observable Particulate Material in the Bulk Solution

Operational Condition	Description or Example
Cell bursting agent	Lauryl sulfate
Colloidal floc	Nondegrading or slowly degrading colloids
Elevated temperature	>32 °C
Foam production	Foam-producing filamentous organisms
Increase in percent MLVSS	Accumulation of fats, oils, and grease
Lack of ciliated protozoa	<100/mL
Low dissolved oxygen concentration	<1 mg/L for 10 consecutive hours
Low pH/high pH	<6.5/>8.5
Nutrient deficiency	Usually nitrogen or phosphorus
Salinity	Excess sodium and/or potassium
Septicity	ORP < −100 mV
Shearing action (excess turbulence)	Surface aeration
Slug discharge of soluble cBOD	3× typical soluble cBOD
Surfactant	Anionic detergents
Total dissolved solids (TDS)	>5000 mg/L
Toxicity	Return activated sludge (RAS) chlorination
Viscous floc or zoogloeal growth	Rapid proliferation of floc-forming bacteria
Young sludge age	>3 days MCRT

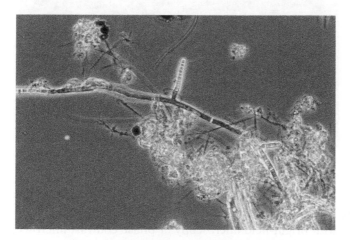

Figure 13.4   *Insignificant particulate material.*

several fields of view at 100× total power of magnification and determining the relative abundance of particulate material according to the following ratings:

"Insignificant" (Figure 13.4)
Most fields of view have little or no particulate material in the bulk solution
"Significant" (Figure 13.5)
Most fields of view have free-floating particulate material in the bulk solution

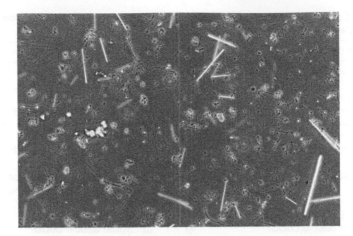

Figure 13.5 *Significant particulate material.*

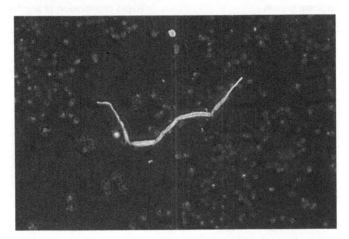

Figure 13.6 *Particulate material under methylene blue stain.*

Particulate material can be observed in a wet mount of mixed liquor at 100× total magnification with bright-field microscopy or phase contrast microscopy. A drop of methylene blue may be added to the wet mount to more easily observe the particulate material under bright-field microscopy (Figure 13.6).

14

Spirochetes

Spirochetes are a group of Gram-negative bacteria. They are slender and long (5–50 μm) with a flexible helical shape (Figure 14.1). They are very motile and move in a corkscrew-shaped motion.

There are aerobic, facultative anaerobic, and anaerobic spirochetes, and they usually proliferate in wastewater or wastewater samples during transitions between aerobic and anaerobic conditions.

The group is highly diverse. Although some spirochetes are pathogenic, for example, causative agents for syphilis, most are free-living and are found in soil. They enter activated sludge processes through inflow and infiltration.

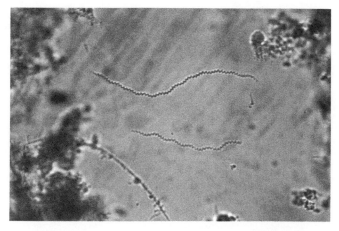

Figure 14.1   *Free-swimming, helical-shaped spirochete.*

Microscopic Examination of the Activated Sludge Process, by Michael H. Gerardi
Copyright © 2008 John Wiley & Sons, Inc.

14

Part IV

Floc Particles and Foam

15

Floc Particles

The activated sludge process is able to efficiently treat wastewater because of the development and maintenance of mature floc particles (Figure 15.1). Bacteria, bacterial secretions, adsorbed fats, oils, grease, colloids, and degradable and nondegradable particulate material are present in floc particles. Bacteria, the major component of floc particles, are present in billions per gram. Bacteria are present in not only large numbers but also large diversity. This enables the activated sludge process to treat a large quantity and large variety of wastes.

Floc formation is achieved through the "stressing" or aging of floc-forming bacteria. These bacteria enter the activated sludge process through fecal waste and I/I. With increasing sludge age or MCRT, floc-forming bacteria produce the necessary cellular components for agglutination (Figure 15.2). These components consist of fibrils, starch granules, and sticky polysaccharides.

FLOC STRUCTURE

The structural properties and characteristics of the majority of the floc particles in an activated sludge process influence treatment efficiency and solids settleability, compaction, and dewaterability. These properties and characteristics also reflect the health of the mixed liquor. A microscopic examination of floc particles can be useful in identifying desirable and undesirable structural properties and characteristics. Often, the examination can reveal the operational conditions responsible for their formation.

Several significant structural properties and characteristics of floc particles can be identified by microscopic examination. These properties and characteristics include (1) shape, (2) size, (3) range in size, (4) color, (5) strength, (6) filamentous

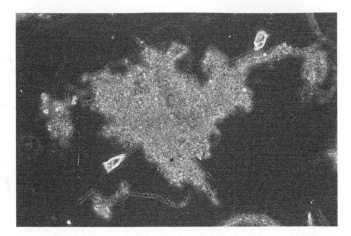

Figure 15.1   *Mature floc particle. In a healthy, steady-state condition a mature floc particle would be golden-brown in color, irregular in shape, and medium (150–500 μm) or large (>500 μm) in size. The floc particle would possess limited filamentous organism growth. The bulk solution would contain little dispersed growth and little particulate material. Crawling ciliates and stalked ciliates may be found on the floc particle.*

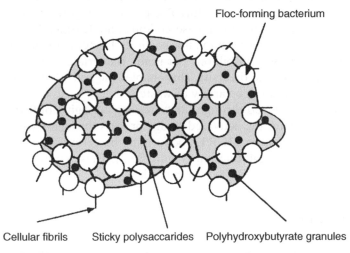

Figure 15.2   *Cellular components necessary for floc formation.*

organism and floc particle structures, (7) relative quantity of stored food, and (8) zoogloeal growth.

SHAPE

Floc particles within an activated sludge process are commonly spherical (Figure 15.3) and irregular (Figure 15.4) in shape. An infrequently observed shape for floc particles is oval, sometimes referred to as congealed (Figure 15.5).

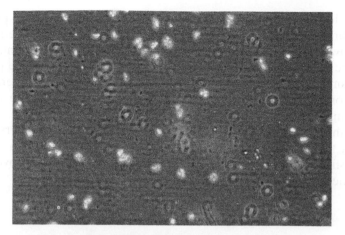

Figure 15.3 *Spherical floc particles.*

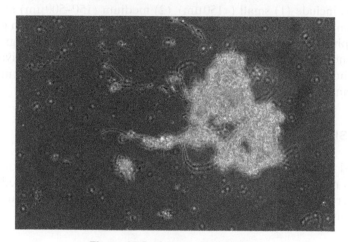

Figure 15.4 *Irregular floc particles.*

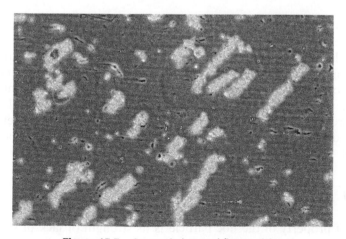

Figure 15.5 *Congealed or oval floc particles.*

Spherical floc particles either lack filamentous organism growth or possess only a small amount of filamentous organism growth. The lack of adequate filamentous organism growth permits shearing of the particle during aeration, mixing, and transferring of floc particles (solids) in the activated sludge process. In the presence of adequate filamentous organism growth, floc bacteria grow along the lengths of the filamentous organisms, resulting in an increase in size and a change in shape to irregular from spherical.

The oval-shaped floc particle or congealed floc particle may be observed in a wet mount under two conditions. These conditions include (1) the use of a dirty or oily microscope slides and (2) the presence of excess heavy metals and overdose of a coagulant (metal salt) or polymer.

SIZE

Floc particles are commonly placed into one of three groups with respect to size. These groups include (1) small (<150 μm), (2) medium (150–500 μm) and (3) large (>500 μm). Small floc particles possess little or no filamentous organism growth, and usually are spherical in shape. Without adequate filamentous organism growth, these particles cannot overcome the turbulence or shearing action in the activated sludge process. Medium and large floc particles typically have adequate filamentous organism growth and usually are irregular in shape.

RANGE IN SIZE

Active mixed liquor usually contains floc particles of all three size groups, and the floc particles may range in size from a few microns to several hundred or several thousand microns. Shifts in the range in sizes of floc particles may indicate significant changes in wastewater strength or composition.

COLOR

The natural color of floc particles is determined by the age of the floc particle or the sludge age of the activated sludge process. Young bacteria produce only a small quantity of oils that darken floc particles. Therefore, young bacteria produce lightly colored or white floc particles. With increasing sludge age, bacteria in the floc particles become older and produce copious quantities of oils that accumulate in floc particles. The accumulation of oils produces a golden-brown color. Therefore, the active and growing floc particle possesses a golden-brown core where most old bacteria are located and a white perimeter where most young bacteria are located (Figure 15.6).

STRENGTH

An important property of the floc particle is its strength or lack of strength. Firm and dense floc particles possess floc bacteria that are tightly adjoined. These parti-

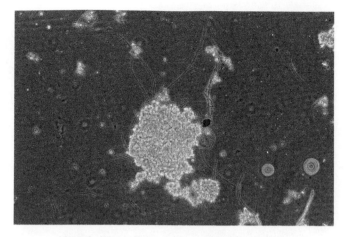

Figure 15.6   *Active and growing floc particles.*

TABLE 15.1   Operational Conditions Associated with the Development of Weak and Buoyant Floc Particles

Cell bursting agent
Elevated temperature
Increase in percent MLVSS
Low dissolved oxygen concentration
Low pH/High pH
Salinity
Septicity
Shearing action
Slug discharge of soluble cBOD
Surfactant
Total dissolved solids
Viscous floc or zoogloeal growth
Young sludge age

cles are tolerant of shearing action and settle or compact nicely in the secondary clarifier.

Weak and buoyant floc particles possess floc bacteria that are loosely adjoined. These particles may be easily sheared and settle poorly in the secondary clarifier. Often, the floc bacteria are separated a great distance by the deposition of exocellular polymeric materials or loosely adjoined because of adverse operational conditions such as swings in pH or presence of surfactants (Table 15.1).

The strength of floc particles may be observed during a microscopic examination of mixed liquor by preparing a wet mount stained with methylene blue. Under methylene blue, floc bacteria stain dark blue (Figure 15.7), while exocellular polymeric materials stain slight blue and openings or voids in the floc particles stain the same intensity of blue as the bulk solution surrounding the floc particles (Figure 15.8).

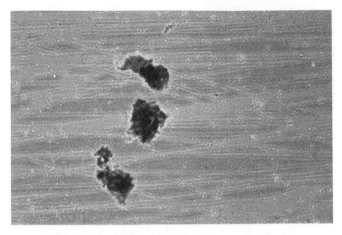

Figure 15.7 Firm floc particles under methylene blue.

Figure 15.8 Weak floc particles under methylene blue.

FILAMENTOUS ORGANISM AND FLOC PARTICLE STRUCTURES

There are two undesired patterns of growth for filamentous organisms and floc particles. These patterns are interfloc bridging and open floc formation. Interfloc bridging is the joining in the bulk solution of the extended filamentous organisms from the perimeter of two or more floc particles (Figure 15.9). Open floc formation is the scattering of the floc bacteria in many small groups along the lengths of the filamentous organisms (Figure 15.10). Significant interfloc bridging and significant open floc formation adversely affect solids settleability in the secondary clarifier.

RELATIVE QUANTITY OF STORED FOOD

During a nutrient deficiency, usually for nitrogen or phosphorus, floc bacteria cannot properly degrade soluble cBOD. The soluble cBOD that is absorbed by the bacteria,

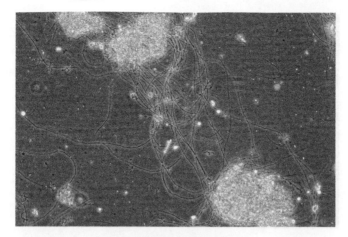

Figure 15.9 *Interfloc bridging.*

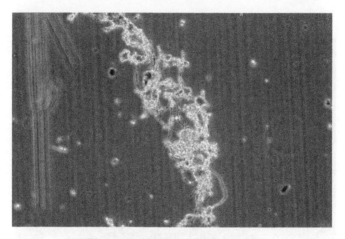

Figure 15.10 *Open floc formation.*

is converted to an insoluble starch and stored in the floc particles between the bacteria. When nutrients become available for bacterial use, the starch is solubilized, absorbed by the bacteria, and then degraded. A nutrient deficiency typically occurs for nitrogen when the mixed liquor effluent filtrate for ammonical-nitrogen $(NH_4^+\text{-}N)$ is <1 mg/L or nitrate-nitrogen $(NO_3^-\text{-}N)$ is <3 mg/L, if ammonical-nitrogen is not present. A nutrient deficiency usually occurs for phosphorus when the mixed liquor effluent filtrate for orthophosphate $(HPO_4^{2-}/H_2PO_4^-)$ or reactive phosphorus is <0.5 mg/L.

Nutrient-deficient floc particles contribute to settleability problems and dewatering problems. The floc particles also produce billowy white foam (young sludge age) or greasy gray foam (old sludge age).

The relative quantity of stored food in floc particles can be observed by performing the India ink reverse stain. Under the india ink reverse stain stored food appears white, while bacterial cells appear black or golden-brown. Although most floc particles possess some stored food (Figure 15.11), these particles respond negatively to

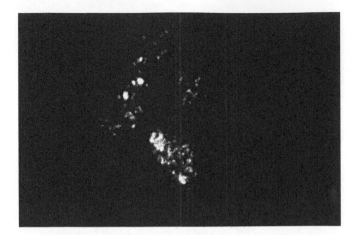

Figure 15.11   Negative india ink reverse stain.

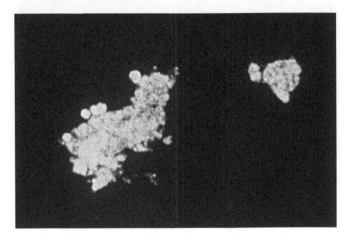

Figure 15.12   Positive india ink reverse stain.

the stain, that is, most of the area of the floc particles is black or golden-brown. Floc particles with significant quantities of stored food respond positively to the stain (Figure 15.12), that is, most of the area of the floc particles is white.

ZOOGLOEAL GROWTH

Operational conditions such as septicity or fermentation upstream of an aeration tank, a nutrient deficiency, long hydraulic retention time (HRT), low pH, high MCRT, and high F/M may trigger zoogloeal growth. Zoogloeal growth may be present in either the amorphous (globular) form (Figure 15.13) or the dendritic (fingerlike) form (Figure 15.14).

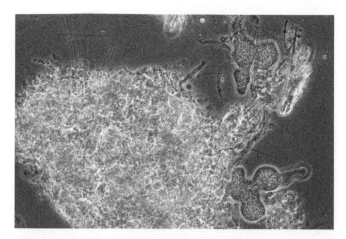

Figure 15.13   *Amorphous or globular zoogloeal growth (upper right and lower right).*

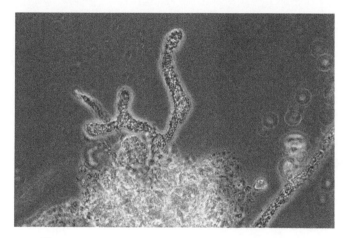

Figure 15.14   *Dendritic or "fingerlike" zoogloeal growth.*

Zoogloeal growth or viscous floc is the rapid and undesired growth of floc-forming bacteria such as *Zoogloea ramigera*. The growth results in the production of weak and buoyant floc particles and may be associated with the production of billowy white foam. Most zoogloeal organisms are strict aerobes and can be controlled with the use of anoxic periods.

Figure 15.13 Amorphous or globular congealed growth (fat).

16

Tetrads

Tetrads or tetrad-forming organisms (TFO) are mostly cyanobacteria (blue-green algae). They are large and spherical, usually grow in clusters of four, and may be referred to as quadricoccus (Figure 16.1). Tetrads grow in most environments where phosphorus is sufficient to support their growth and are found in large numbers in activated sludge processes that practice biological phosphorus removal ("luxury uptake" of phosphorus).

Although they can degrade simple organic compounds, tetrads are capable of using carbon dioxide (CO_2) as their carbon source. They can fix molecular nitrogen

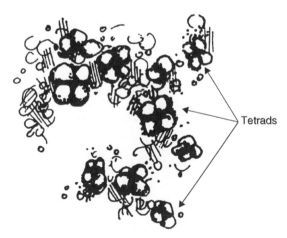

Tetrads

Figure 16.1 *Tetrads. Tetrads consist of four spherical cyanobacteria or blue-green algae. They usually are found on the surface of floc particles and stain dark red (negative) under the Gram stain.*

Microscopic Examination of the Activated Sludge Process, by Michael H. Gerardi
Copyright © 2008 John Wiley & Sons, Inc.

(N_2) as their nitrogen source when nitrogen is deficient in wastewater. Because they can fix molecular nitrogen, tetrads often bloom in wastewater treatment systems during a nutrient deficiency for nitrogen.

Tetrads vary greatly in shape and appearance. In addition to clusters of four, cyanobacteria may be found as unicellular, colonial, and filamentous organisms. Filamentous cyanobacteria have a gliding motion. They appear similar to yeast and algae and stain Gram-negative and Neisser-positive.

Tetrad-forming organisms reproduce by binary fission, budding, and multiple fission. Rapid proliferation of the organisms typically is due to a nutrient deficiency for nitrogen and atypically is due to high BOD loading. Tetrads settle poorly. Therefore, proliferation of tetrads results in poor settleability of solids and high total suspended solids (TSS) in the final effluent.

Undesired numbers of tetrads are found in paper mill lagoons more often than municipal activated sludge processes. These lagoons commonly experience a nutrient deficiency for nitrogen and have warm wastewater temperatures. Thermophilic species of cyanobacteria can grow at temperatures up to 75 °C. Increasing the soluble nitrogen concentration in the influent and reducing the BOD loading may achieve control of undesired numbers of tetrads.

17

Zoogloeal Growth

Zoogloeal growth or viscous floc is the rapid and undesired proliferation of floc-forming bacteria (Table 17.1). The growth is termed "zoogloeal" after the genus name of the first floc-forming bacterium, *Zoogloea ramigera*, that was identified as growing in a problematic form in the activated sludge process. This bacterium is a rod-shaped (0.5–1.0 μm × 1.0–3.0 μm), Gram-negative organotroph that produces large quantities of gelatinous exocellular polysaccharides. The polysaccharides are less dense than wastewater, hinder compaction of the floc bacteria, and entrap air and gas bubbles.

Although floc-forming bacteria perform two significant and beneficial roles in the activated sludge process—floc formation and degradation of cBOD—their rapid proliferation often results in the production of weak and buoyant floc particles, settleability problems, loss of solids from the secondary clarifier, and production of billowy white foam.

Excessive zoogloeal growth is found in two patterns, amorphous (lacking specific shape) or globular (Figure 17.1) and dendritic or "fingerlike" projections (Figure 17.2). Zoogloeal growth may also appear as a white or gray-white slimy film on the walls or weirs of secondary clarifiers. Zoogloeal growth may enter the activated

TABLE 17.1 Significant Genera of Floc-Forming Bacteria

Achromobacter	*Citromonas*
Aerobacter	*Escherichia*
Alcaligenes	*Flavobacterium*
Arzthrobacter	*Pseudomonas*
Bacillus	*Zoogloea*

Microscopic Examination of the Activated Sludge Process, by Michael H. Gerardi
Copyright © 2008 John Wiley & Sons, Inc.

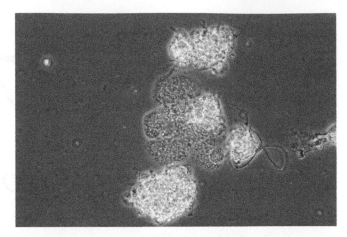

Figure 17.1 *Amorphous or globular zoogloeal growth.*

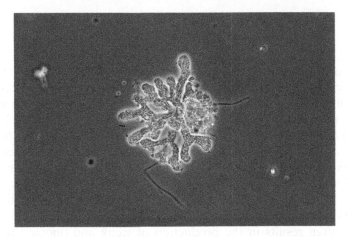

Figure 17.2 *Dendritic or "fingerlike" zoogloeal growth.*

sludge process from the sloughing of biofilm from fixed processes that are located upstream of the activated sludge process.

Operational conditions associated with zoogloeal growth include nutrient deficiency, long HRT, low pH, high MCRT, and high F/M. The presence of volatile fatty acids from selector systems and septic wastewater also triggers the undesired growth of zoogloeal organisms. Zoogloeal growth may be controlled with the use of an appropriate polymer or exposure of the growth to anoxic periods of 1–2 hours. Because of the copious production of gelatinous exocellular polysaccharides, an anionic polymer may better capture and thicken zoogloeal organisms.

18

Foam

Foam (Figure 18.1) is entrapped air and/or gas bubbles beneath a layer of solids. Biological foam is produced in the aeration tank and often escapes to the secondary clarifier and other wastewater treatment tanks. When foam moves from tank to tank or falls over weirs, the entrapped air and gas bubbles escape and the foam collapses. Collapsed foam often is referred to as scum (Figure 18.2).

There are six biological conditions that produce foam in the activated sludge process (Table 18.1). Foam may be described by specific texture and color, for example, viscous and chocolate-brown for foam-producing filamentous organisms. Biological conditions that produce foam include (1) undesired growth of foam-producing filamentous organisms, (2) nutrient deficiency at an old sludge age, (3) nutrient deficiency at a young sludge age, (4) undesired zoogloeal growth, (5) slug discharge of soluble cBOD, and (6) erratic wasting rates for secondary solids.

The occurrence of operational conditions responsible for foam production can be suspected by physical observations of the aeration tank and other wastewater treatment tanks. Microscopic examinations of wet mounts and smears of the mixed liquor or stained smears of foam can identify the operational conditions responsible for most biological foam production (Table 18.2).

FOAM-PRODUCING FILAMENTOUS ORGANISMS

There are three known foam-producing filamentous organisms. These organisms are *Microthrix parvicella* (Figure 18.3), nocardioforms (Figure 18.4), and type 1863 (Figure 18.5). *Microthrix parvicella* and nocardioforms are responsible for most occurrences of foam by foam-producing filamentous organisms and are easily

Microscopic Examination of the Activated Sludge Process, by Michael H. Gerardi
Copyright © 2008 John Wiley & Sons, Inc.

Figure 18.1   *Foam. Foam such as nocardial foam is entrapped air and/or gas bubbles beneath a layer of solids. The layer of solids in nocardial foam consists of lipids released by nocardioforms in floc particles.*

Figure 18.2   *Collapsed foam or scum. When entrapped air and/or gas bubbles escape beneath a layer of solids, the solids collapse and the collapsed foam often is referred to as scum.*

TABLE 18.1   Operational Conditions Responsible for Biological Foam Production

Operational Condition	Foam
Foam-producing filamentous organisms	Viscous chocolate-brown
Nutrient deficiency, old sludge age	Greasy gray
Nutrient deficiency, young sludge age	Billowy white
Zoogloeal growth	Billowy white
Slug discharge of soluble cBOD	Billowy white
Erratic wasting rates for secondary solids	Concentric circles of light brown and dark brown

TABLE 18.2   Microscopic Observations Related to Biological Foam Production

Foam	Slide Preparation	Stain	Confirmation
Foam-producing filamentous organisms	Mixed liquor smear	Gram	Undesired growth of *Microthrix parvicella*, nocardioforms, type 1863
	Foam smear	Gram	Undesired growth of *Microthrix parvicella*, nocardioforms, type 1863
Nutrient deficiency	Mixed liquor wet mount	India ink	Numerous positive reactions to India ink reverse stain
Zoogloeal growth	Mixed liquor wet mount	Methylene blue	Significant amorphous or dendritic zoogloeal growth
Slug discharge of soluble cBOD	Mixed liquor smear	Safranin	Numerous floc particles with firm core and weak perimeter

Figure 18.3   Microthrix parvicella. Microthrix parvicella *is a Gram-positive filamentous organism and appears dark in the photomicrograph. Foam produced by this filamentous organism is viscous and chocolate-brown. Under bright-field microscopy the Gram-positive filamentous organism often appears as a string of blue "beads."*

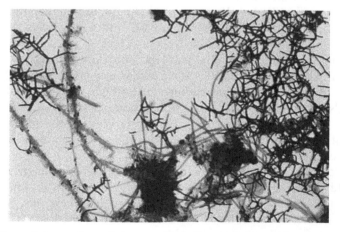

Figure 18.4   *Nocardioforms.* Nocardioforms *or* Nocardia *are Gram-positive filamentous organisms that produce viscous, chocolate-brown foam.*

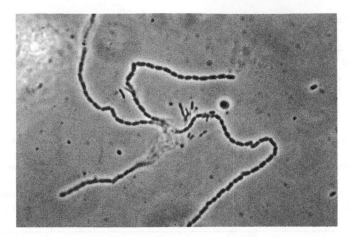

Figure 18.5 *Type 1863. Type 1863 is a Gram-negative foam-producing filamentous organism. The bacterial cells in the filamentous organism are bacillus or rod-shaped with constrictions, that is, the ends of the cells are pinched in like "sausage links."*

identified through microscopic examinations of smears of mixed liquor and smears of foam. These filamentous organisms typically are present in foam in much higher abundance than mixed liquor. Although numerous filamentous organisms may be found in any foam, this does not mean that foam production was caused by the undesired growth of foam-producing filamentous organisms. Care should be taken to ensure that foam-producing filamentous organisms are identified in foam.

NUTRIENT DEFICIENCY

A nutrient deficiency can be detected by positive reactions of numerous floc particles to India ink reverse stain using a wet mount of mixed liquor. This stain reveals the relative amount of stored food in floc particles. If stored food is small in quantity, the test result is considered to be negative, and the probability of a nutrient deficiency is considered to be low. If stored food is large in quantity, the test result is considered to be positive, and the probability of a nutrient deficiency is considered to be high.

During a nutrient deficiency, floc bacteria store large quantities of substrate (cBOD) as insoluble polysaccharides (exocellular polymeric material) within floc particles. The polysaccharides that capture air and gas bubbles resulting in foam production also block the movement of the carbon black particles in the India ink as they move through the floc particles. The larger the quantity of polysaccharides that is present in the floc particle, the larger the area of white or nonstained polysaccharides there is in the floc particles.

Nutrient-deficient foam may be billowy white at a young sludge age or greasy gray at an old sludge age. At an old sludge age numerous oils that have been produced by aging bacteria accumulate in the floc particles. These oils darken the floc particles to a golden-brown color and are transferred to the foam. The transferred

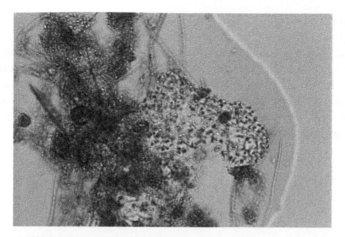

Figure 18.6   *Zoogloeal growth. The rapid growth of young, floc-forming bacteria such as* Zoogloea ramigera *results in the production of a large quantity of gelatinous material. This material separates the cells a great distance. Under the Gram stain the quantity of gelatinous material can be seen as the clear space between the cells. Zoogloeal growth results in the production of weak and buoyant floc particles. When air and/or gas bubbles are captured in the gelatinous materials, billowy white foam may be produced.*

oils darken foam to a greasy gray. At a young sludge age very few oils are produced and accumulated in floc particles. Therefore, the floc particles are white in color, and foam is billowy white as few oils are transferred to foam.

ZOOGLOEAL GROWTH

Zoogloeal growth or viscous floc is the rapid and undesired growth of floc-forming bacteria. The growth results in the production and accumulation of a copious quantity of exocellular gelatinous material that captures air and gas bubbles.

Confirmation of billowy white foam produced by zoogloeal growth can be obtained only by the detection of significant quantities of its amorphous and dendritic patterns of proliferation. This can be observed in wet mounts of mixed liquor. To more easily observe zoogloeal growth a drop of methylene blue may be added to the wet mount. A Gram-stained smear of mixed liquor can reveal the relatively large quantity of gelatinous materials between the bacterial cells (Figure 18.6).

SLUG DISCHARGE OF SOLUBLE CBOD

A slug discharge of soluble cBOD occurs when two to three times the normal quantity of soluble cBOD enters an activated sludge process over a 3- to 4-hour period. The slug discharge results in the rapid growth of floc bacteria and the production of a copious quantity of insoluble polysaccharides by a young bacterial population. The polysaccharides capture air and gas bubbles, which results in the production of billowy white foam in the aeration tank.

The occurrence of a slug discharge of soluble cBOD can be detected microscopically by the presence of numerous floc particles with a firm core and a weak perimeter as revealed by a safranin stain of a smear of mixed liquor. Old bacteria in the core of the floc particles appear dark red because of their tight compaction and the presence of a relatively small quantity of polysaccharides between them. However, rapidly growing young bacteria in the perimeter of the floc particles appear light red because of their poor compaction and the presence of a relatively large quantity of polysaccharides that separate them a large distance.

Figure 18.7 *Pockets of young and old bacterial growth. Erratic wasting rates often result in the production of pockets of young and old bacterial growth that produce different foams. Terminating aeration and mixing action permits the development of concentric circles of light foam (young growth) and dark foam (old growth).*

ERRATIC WASTING RATES FOR SECONDARY SOLIDS

Erratic wasting rates for secondary sludge results in the production of "pockets" of young bacterial growth and old bacterial growth. The pockets of young growth produce lightly colored brown foam, while the pockets of old growth produce dark-colored brown foam. Different color circles of foam can be observed on the aeration tank when aeration and mixing action are terminated (Figure 18.7).

ERRATIC WASTING RATES FOR SECONDARY SOLIDS

Part V

Protozoa

19

Protozoa

Protozoa are single-celled organisms (Figure 19.1). However, some are colonial (Figure 19.2). Most protozoa range in size from 5 to 250 μm and are predominantly saprozoic. Some are holophytic.

PROTOZOAN GROUPS

There are six groups of protozoa that are commonly observed in the activated sludge process. These groups are commonly identified and referred to as simple or "lower" life-forms to complex or "higher" life-forms. **Amoebae** (Figure 19.3) are referred to as the simplest life-forms and have a cytoplasm or "gut" content that flows against a very flexible cell membrane. This flowing action permits locomotion. There are two types of amoebae—the naked and the testate. The testate amoebae have a protective covering or shell. Amoebae move slowly in the bulk solution or "drift" in water currents. **Flagellates** (Figure 19.4) are typically small and oval in shape. They possess whiplike structures or flagella that beat to provide locomotion. The beating action of the flagella provides a "corkscrew" pattern of locomotion.

Free-swimming ciliates (Figure 19.5) have short hairlike structures or cilia in rows on the entire cellular surface. The cilia beat in unison and provide locomotion. The beating action of the cilia provides a fairly straight line of locomotion in the bulk solution. **Crawling ciliates** (Figure 19.6) have rows of cilia only on the ventral surface of the cell and prefer to be attached to floc particles. The beating action of the cilia provides locomotion and produces water currents that draw bacteria to their mouth opening on the ventral surface. Crawling ciliates possess modified cilia or cirri that

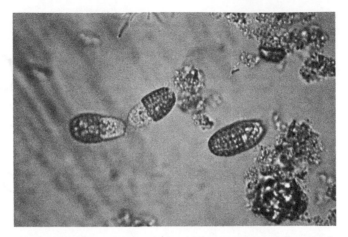

Figure 19.1 Solitary protozoa. Coleps are free-swimming ciliates. They also have a testate.

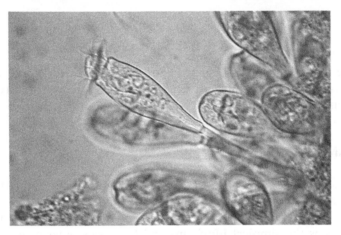

Figure 19.2 Colonial protozoa. Opercularia is a colonial protozoa.

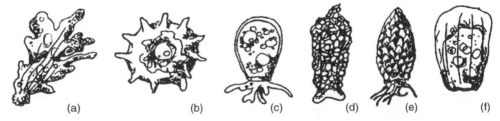

(a)   (b)   (c)   (d)   (e)   (f)

Figure 19.3 Amoebae. Commonly observed amoebae in the activated sludge process include Amoeba (a), Arcella (b), Cryptodifflugia (c), Difflugia (d), Euglypha (e), and Thecamoeba (f).

appear as "spikes" and help to anchor the organism to the surface of floc particles.

Stalked ciliates (Figure 19.7) are referred to as the highest life-forms, possess cilia only around the mouth opening, and prefer to be attached to the floc particles. The beating action of the cilia produces water currents that draw bacteria into the mouth

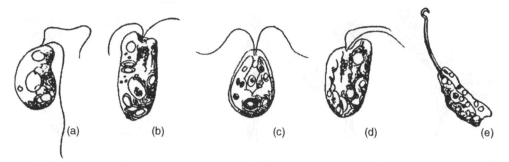

Figure 19.4 *Flagellates. Commonly observed flagellates in the activated sludge process include* Bodo (a), Chilomonas (b), Chlamydomonas (c), Cryptomonas (d), *and* Peranema (e).

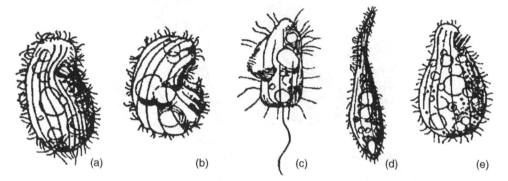

Figure 19.5 *Free-swimming ciliates. Commonly observed free-swimming ciliates in the activated sludge process include* Colpidium (a), Colpoda (b), Cyclidium (c), Litonotus (d), *and* Tetrahymena (e).

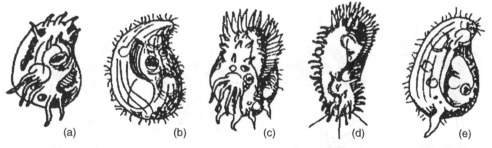

Figure 19.6 *Crawling ciliates. Commonly observed crawling ciliates in the activated sludge process include* Aspidisca (a), Chilodonella (b), Euplotes (c), Stylonychia (d), *and* Trochilia (e).

opening. Some stalked ciliates such as *Zoothamnium* (Figure 19.8) have a contractile filament in the stalk that permits the organism to "spring." The springing action also produces a water vortex to draw bacteria into the mouth opening. The contractile filament is not present in some stalked ciliates such as *Opercularia* (Figure 19.7). An interesting and odd group of protozoa are the **Suctoria** (Figure 19.9). These organisms possess tentacles rather than cilia.

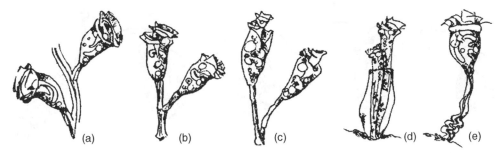

Figure 19.7 *Stalked ciliates. Commonly observed stalked ciliates in the activated sludge process include* Carchesium (a), Epistylis (b), Opercularia (c), Vaginicola (d), *and* Vorticella (e).

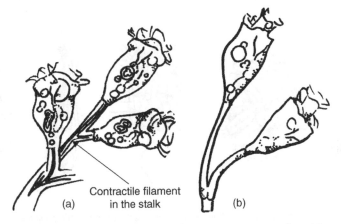

Figure 19.8 *Contractile filament. The contractile filament or myoneme is not found on all stalked ciliates.* Zoothamnium (a) *has a contractile filament and can "spring," while* Opercularia (b) *does not have a contractile filament and cannot "spring."*

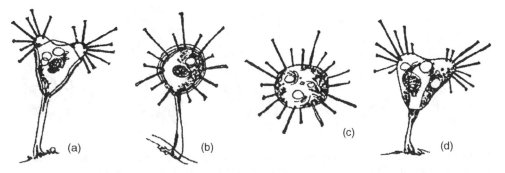

Figure 19.9 *Suctoria. Commonly observed suctoria or tentacled ciliates in the activated sludge process include* Acineta (a), Podophrya (b), Sphaerophrya (c), *and* Tokophrya (d).

PROTOZOA AS BIOINDICATORS

Protozoa, especially ciliated protozoa, perform several important roles in the treatment of wastes in the activated sludge process. These roles include the removal of fine solids—colloids, dispersed growth, and particulate material—from the bulk solution, recycling of mineral nutrients, and promotion of floc formation. In addition to these roles, many wastewater treatment plant operators use protozoa as an indicators or bioindicators of the health of the activated sludge process and the effluent quality of the mixed liquor. Protozoa indicate the quality of the mixed liquor effluent. Mixed liquor protozoa do not indicate bulking solids in the secondary clarifier and the loss of these solids from the activated sludge process. There are two methods that may be used to determine the health of the activated sludge and its effluent quality.

First, microscopic scans of wet mounts of mixed liquor can be examined to determine the dominant groups of protozoa. Higher life-forms (stalked ciliates and crawling ciliates) generally indicate healthy mixed liquor and acceptable mixed liquor effluent quality, while lower life-forms (flagellates and amoebae) generally indicate unhealthy mixed liquor and unacceptable mixed liquor effluent quality. This method is simple and quick, but the method may be misleading at times, because there are higher life-forms that proliferate under poor operational conditions and, conversely, there are lower life-forms that proliferate under excellent operational conditions (Table 19.1).

There are four operational conditions that affect the relative abundance of specific protozoa in the activated sludge process. These conditions include (1) dissolved oxygen concentration, (2) flow, (3) MCRT, and (4) organic loading. The relationship between these conditions can be observed in the following:

- Rapid flow rates favor organisms with short generation times and a small diversity of life-forms (small flagellates and small ciliates).

TABLE 19.1 Examples of Ciliated Protozoa as Bioindicators of Mixed Liquor Effluent Quality

Stalked ciliates
  Poor-quality effluent (for example, BOD > 20 mg/L)
    Epistylis plicatilis
    Opercularia coaractata
    Vorticella alba
  Good-quality effluent (for example, <20 mg/L)
    Epistylis rotans
    Vorticella elongate
    Zoothamnium mucedo

Crawling ciliates and tentacled, stalked ciliates
  Poor-quality effluent
    Podophrya fixa
    Sphaerophyra magna
  Good-quality effluent
    Euplotes affinis
    Podophrya maupasi

Free-swimming ciliates
  Poor-quality effluent
    Paramecium trichium
    Trachelphyllum pusillum
  Good-quality effluent
    Paramecium aurelia
    Litonotus anguilla

Figure 19.10 Saprobic index and indicator organisms. The saprobic index as applied to the activated sludge process has four conditions—polysaprobic, alpha-mesoprobic, beta-mesoprobic, and oligosaprobic. The polysaprobic condition is characterized by high flow, severe organic overload, an anaerobic or low dissolved oxygen condition, very poor biota health, turbid and poor mixed liquor effluent supernatant, and high concentrations of ammonia and sulfide. Bioindicator organisms of a polysaprobic condition include Amoeba (a), Bodo caudatus (b), Euglypha (c), Peranema trichophorum (d), Tetrahymena pyriformis (e), Trepomonas agilis (f), and Vampyrella (g).

The alpha-mesoprobic condition is characterized by high flow, organic overload, low dissolved oxygen concentration, poor biota health, poor mixed liquor effluent supernatant, and presence of significant concentrations of ammonia and sulfides. Bioindicator organisms of an alpha-mesoprobic condition include Chilodonella uncinata (h), Difflugia (i), Hexamitus fissus (j), Litonotus fasciola (k), Pleuromonas jaculans (l), Podophrya fixa (m), and Vorticella convallaria (n).

The beta-mesoprobic condition is characterized by low flow, high organic loading, adequate dissolved oxygen condition, moderate biota health, acceptable mixed liquor effluent supernatant, and presence of nitrate (nitrification) and sulfate (sulfide oxidation). Bioindicator organisms include Arcella (o), Aspidisca costata (p), Euplotes affinis (q), Opercularia microdiscum (r), Tokophrya quadripartite (s), Trachelophyllum pusillum (t), and Vorticella alba (u).

The oligosaprobic condition is characterized by low flow, moderate organic loading, adequate dissolved oxygen condition, excellent biota health, "polished" mixed liquor effluent supernatant, and presence of significant concentrations of nitrate and sulfate. Bio-indicator organisms include Epistylis rotans (v), nematode (w), rotifer (x), Stylonchia pustulata (y), Vorticella nebulifera (z), and Vorticella striata (aa).

- Slow flow rates favor organisms with short and long generation times and a large diversity of life-forms (crawling ciliates, stalked ciliates, rotifers, and free-living nematodes).
- High organic loadings produce low dissolved oxygen concentrations that favor protozoa that can survive anaerobic conditions (amoebae, flagellates, and small ciliates).
- Low organic loadings produce high dissolved oxygen concentrations that favor organisms that are strict aerobes (crawling ciliates, stalked ciliates, suctoria, rotifers, and free-living nematodes).
- Low organic loadings and high dissolved oxygen concentrations favor a large diversity of life-forms.

Therefore, a second or more detailed examination or identification of dominant protozoa to their genus name such as *Difflugia* or to their scientific name (genus and species) such as *Euplotes affinis* is required. This method utilizes the "saprobic" index.

The saprobic index or *Saprobiensystem* is European and is used to classify organisms according to their response to the organic pollution in slow-moving streams. The index may be modified to describe four operational conditions in an activated sludge process (Figure 19.10). These conditions include (1) polysaprobic (grossly polluted—contains the complex organic wastes that are decomposing primarily by anaerobic processes), (2) alpha-mesosaprobic (polluted—significant organic wastes that are decomposing by anaerobic and aerobic processes), (3) beta-mesosaprobic (moderately polluted—organic wastes that are decomposing by aerobic processes), and (4) oligosaprobic (slightly polluted—none of the original organic wastes remain, organic wastes produced through self-purification). Each condition is described in

Flow, decreasing ⟶
Organic loading, decreasing ⟶
Dissolved oxygen concentration, increasing ⟶
MCRT, increasing ⟶

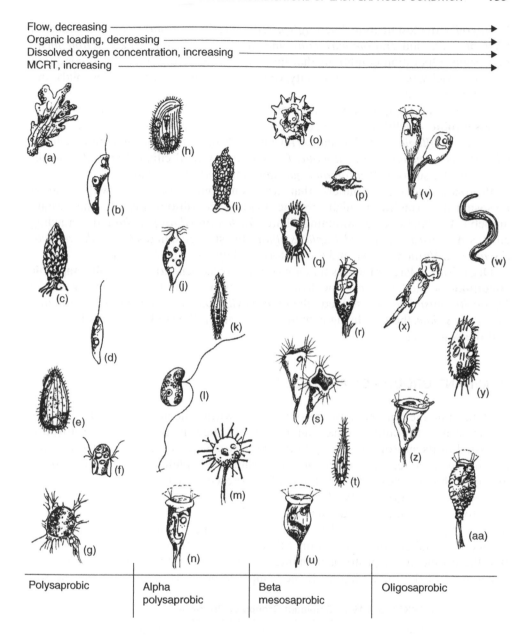

Polysaprobic	Alpha polysaprobic	Beta mesosaprobic	Oligosaprobic

Figure 19.10 in terms of relative flow, organic loading, dissolved oxygen concentration, representative organisms, and mixed liquor effluent supernatant quality.

PROTOZOA AS BIOINDICATORS OF EACH SAPROBIC CONDITION

Protozoa in large numbers that are commonly associated with a **polysaprobic condition**—a severely overloaded organic condition or operational condition that mimics a severely overloaded organic condition—include the naked amoebae *Amoeba* and

Vampyrella, the testate amoebae *Euglypha*, the flagellates *Bodo caudatus, Peranema trichophorum*, and *Trepomonas agilis*, and the free-swimming ciliate *Tetrahymena pyriformis*, Operation conditions that may mimic a severely overloaded condition include toxicity, recovery from toxicity, young sludge age, overwasting of solids, and hydraulic washout.

Protozoa in large numbers that are commonly associated with an **alpha-mesosaprobic condition**—an overloaded organic condition—include the testate amoebae *Difflugia*, the flagellate *Hexamitus fissus*, the free-swimming ciliates *Chilodonella uncinata* and *Litonotus fasciola*, the stalked ciliate *Vorticella convallaria*, and the suctoria *Pleuromonas jaculans* and *Podophrya fixa*.

Protozoa in large numbers that are commonly associated with a **beta-mesosaprobic condition**—high organic loading condition—include the testate amoeba *Arcella*, the free-swimming ciliate *Trachelophyllum pusillum*, the crawling ciliates *Aspidisca costata* and *Euplotes affinis*, the stalked ciliates *Opercularia micodiscum* and *Vorticella alba*, and the suctoria *Tokophrya quadripartite*.

Organisms in large numbers that are commonly associated with an **oligosaprobic condition**—a moderate organic loading condition—include the crawling ciliate *Stylonchia pustulata*, the stalked ciliates *Epistylis rotans, Vorticella nebulifera*, and *Vorticella striate*, the free-living nematode, the rotifer, the stalked ciliates *Vorticella*, and the water bear.

DOMINANT GROUPS OF PROTOZOA

To determine the dominant groups of protozoa within the activated sludge process, a scan of a wet mount of mixed liquor is performed. Each observed protozoa is recorded in its appropriate group until 100 protozoa have been recorded or, time permitting, three separate wet mounts are scanned completely. The three protozoan groups with the largest number of individuals (percentage or percent of the profile) are the dominant groups (Worksheet 19.1).

Before a wet mount for dominant group determination, is prepared the mixed liquor sample should be stirred and aerated with a pipette and pipetting bulb before an aliquot is taken for examination. Stirring and aeration are required to ensure that the profile of the protozoan community observed and recorded is indeed representative of the mixed liquor. If the sample is not stirred and aerated, the wet

WORKSHEET 19.1   Dominant Groups of Protozoa

Group	Count	% Composition
Amoebae		
Flagellates		
Free-swimming ciliates		
Crawling ciliates		
Stalked ciliates		
Total count		
Stalk Ciliated Protozoan Activity and Structure		
% Free Swimming	% Forming Bubbles	% Sheared

mount may not contain representative numbers of (1) amoebae, since they cling to the surface of glass or plastic in sampling bottles, (2) flagellates and free-swimming ciliates that are associated mostly with the liquid medium or supernatant in a settled sample of mixed liquor, and (3) crawling and stalked ciliates that are associated mostly with floc particles that are found at the bottom of a settled sample of mixed liquor.

The protozoan scan can be performed with a bright-field or phase contrast microscope at 100× total power magnification. The use of 400× total power magnification may be required to identify some protozoa to their appropriate group. If the protozoa move too rapidly for identification purposes, adding a drop of an immobilizing agent such as Detain™, Protosol™, 1% nickel sulfate, or 1% mercuric chloride to the wet mount can slow their movement. However, if an immobilizing agent is added to the wet mount, a protozoan count cannot be performed because the volume of mixed liquor beneath the coverslip cannot be estimated. The use of an immobilizing agent should be performed ahead of the count in order to identify protozoa that are observed during the count.

PROTOZOAN ACTIVITY AND STRUCTURE

Protozoa in the presence of toxic or inhibitory wastes become sluggish or inactive. The decrease in activity or termination of activity is caused by the attack of the toxic or inhibitory wastes on the enzyme systems or critical structural components of the protozoa. Therefore, an attack of toxic or inhibitory wastes on an activated sludge process can be observed with a microscopic examination of protozoan activity.

During a scan of a wet mount of a toxic-laden sample, the general activity pattern of all protozoan groups should be reviewed and compared with normal activity of a non-toxic-laden sample, that is, a comparison should be made between the activity of amoebae, flagellates, and ciliates in each sample.

Indicator organisms that often are used to monitor activity within the protozoan community are the crawling ciliate *Aspidisca* (Figure 19.11) and the stalked ciliate

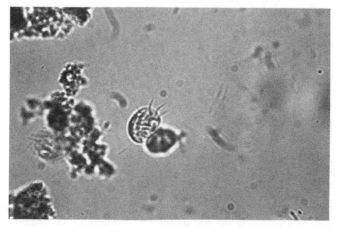

Figure 19.11 Aspidisca.

Vorticella (Figure 19.12). *Aspidisca* are very sensitive protozoa to toxic or inhibitory wastes. In addition *Aspidisca* are easily identified. The beating of the cilia (Figure 19.13) and the springing action of the contractile filament (Figure 19.14) of *Vorticella* also are easily rendered sluggish or inactive in the presence of toxic or inhibitory

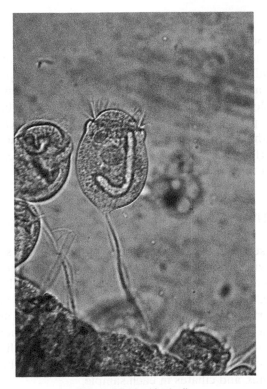

Figure 19.12 Vorticella.

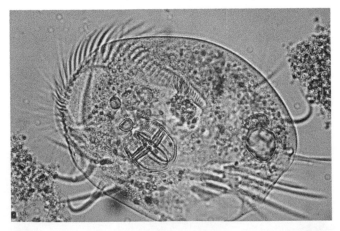

Figure 19.13 *Euplotes. Cilia are present on the anterior surface of the protozoa, while modified cilia or cirri that anchor the protozoa to the floc particle are present on the posterior surface of the protozoa.*

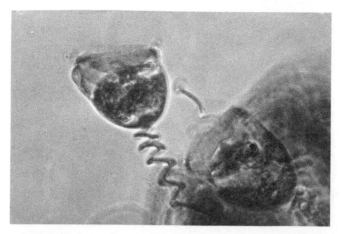

Figure 19.14   Contractile filament. The contractile filament or myoneme in Vorticella is shown contracted under methylene blue.

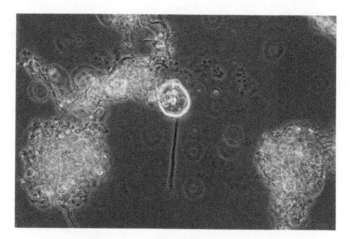

Figure 19.15   Free-swimming stalked ciliate Vorticella.

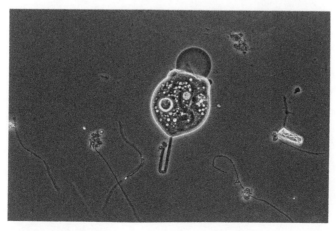

Figure 19.16   Stalked ciliate with bubble leaving the mouth opening. Gas bubble is present in the mouth opening as well as many small gas bubbles in the cytoplasm.

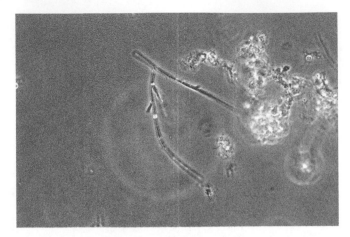

Figure 19.17 Posterior portions or stalks.

wastes. Free-swimming stalked ciliates are indicative of low dissolved oxygen concentration (Figure 19.15). The presence of excessive numbers of "bubble-forming" stalked ciliates may represent severe adverse conditions including toxicity (Figure 19.16). Also, the presence of excessive numbers of sheared stalked ciliates (Figure 19.17) may represent excessive turbulence in the treatment process.

20

Relative Predominance of Bacteria and Protozoa

As an activated sludge process ages without operational problems from start-up to the development of mature floc particles, two significant biological events occur in the mixed liquor. These events involve the bacterial and protozoan communities and are reflective of changes in the degree of pollution of the mixed liquor (Figure 20.1).

As the activated sludge process ages, there occur changes in the population size of bacteria and the numbers of bacteria that are dispersed and flocculated in the mixed liquor. Also, as the activated sludge process ages there occur changes in the relative abundance of protozoa and succession of dominant protozoan groups. The succession of protozoan groups is from amoebae to flagellates to free-swimming ciliates to crawling ciliates to stalked ciliates.

Increase in the relative number of bacteria in the mixed liquor is due to their rapid or exponential growth over time as they degrade cBOD. The maximum size of the bacterial population is limited by the quantity of available cBOD and their consumption by predators, mostly protozoa.

Changes in the relative number of bacteria that are dispersed or flocculated occur through the aging process and the coating action of higher life-forms, mostly protozoa. Most young bacteria are highly motile and remain dispersed in the mixed liquor through the beating action of flagella. As the bacteria age, they produce three necessary cellular components that enable them to "stick" together or flocculate. These components are sticky sugars (insoluble polysaccharides), short fibrils, and starch granules. Many of the starch granules can be observed through PHB staining of a smear of mixed liquor (Figure 20.2).

Changes in the number of protozoa and succession in dominant groups in the mixed liquor are determined by several operational conditions. These conditions

Microscopic Examination of the Activated Sludge Process, by Michael H. Gerardi
Copyright © 2008 John Wiley & Sons, Inc.

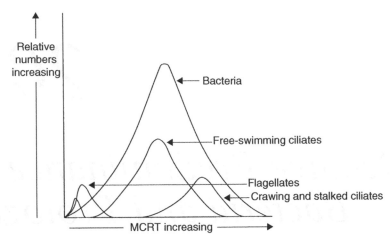

Figure 20.1 *Relative abundance of bacteria and succession of dominant protozoan groups. With increasing mean cell residence time (MCRT), there occurs a change in the relative numbers of bacteria and dominant protozoan groups. These changes in numbers and dominant protozoan groups are caused by decrease in pollution, increase in aeration, change in number of dispersed and flocculate bacteria, and differences in the feeding mechanisms of the different protozoan groups.*

Figure 20.2 *Starch granules. Dark starch granules or poly-β-hydroxybutyrate (PHB) can be seen in the filamentous organism and within the floc particle.*

include organic loading, HRT, dissolved oxygen concentration, MCRT, and substrate availability.

Bacteria obtain substrate as soluble, colloidal, and particulate cBOD. Soluble cBOD is available as dissolved organic compounds such as acids, alcohols, and sugars in the mixed liquor. However, protozoa must compete with bacteria in order to obtain soluble cBOD. Because bacteria are much more numerous in the mixed liquor than protozoa and bacteria have a larger surface-to-volume ratio, they are more efficient in obtaining soluble cBOD than protozoa. Therefore, in the mixed liquor soluble cBOD is not the major source of substrate for protozoa. For protozoa

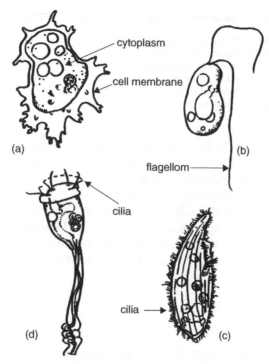

Figure 20.3 *Protozoan locomotion. The streaming action of cytoplasm or gut content against the cell membrane provides locomotion for amoebae (a). The beating action of whiplike structures or flagella provides locomotion for flagellates (b). The beating action of rows of cilia on the entire surface of free-swimming ciliates (c) and around the mouth opening of stalked ciliates (d) provides locomotion for ciliates.*

to successfully compete for soluble cBOD and survive in the mixed liquor, the soluble cBOD in the mixed liquor must be approximately 50,000 mg/L.

Particulate cBOD is available to protozoa in the form of dispersed bacteria. Protozoa graze upon dispersed bacteria. They do not have the ability to remove bacteria from a flocculated mass or floc particle, that is, they cannot burrow into floc particles in order to feed upon bacteria. Therefore, the mode of locomotion and feeding mechanisms used by protozoa to capture dispersed bacteria and the relative abundance of dispersed bacteria greatly determine the dominant protozoan groups in the mixed liquor (Figure 20.3).

AMOEBAE

Amoebae move slowly through the mixed liquor by drifting in water current or by the use of pseudopodia. Amoebae are found in the bulk solution, where they feed upon highly motile, dispersed bacteria. Here, they engulf bacteria within the extended pseudopodia and digest the captured bacteria or particulate cBOD. Pseudopodia movement is not only slow but also inefficient as a food gathering mechanism. However, pseudopodia movement does not require much energy. Also, amoebae are more tolerant of harsh operational conditions (low dissolved oxygen

TABLE 20.1   **Operational Conditions that Mimic a Young Sludge Age**

Hydraulic washout from inflow and infiltration
Low dissolved oxygen concentration
Organic overload or slug discharge or soluble cBOD
Overwasting of mixed liquor suspended solids
Short hydraulic retention time
Toxicity and recovery from toxicity

and high pollution or BOD) than other protozoan groups and have little protozoan competition for dispersed bacteria in harsh operational conditions.

So, with a small energy (substrate) requirement and lack of competition, amoebae are the dominant protozoan group when the bacterial population is relatively small and the mixed liquor is highly polluted. These conditions are reflective of a young sludge age. Although a young sludge age or low MCRT would provide for a large number of dispersed bacteria, there are several operational conditions that would mimic a young sludge age and provide for a large number of dispersed bacteria (Table 20.1).

FLAGELLATES

Flagellates move quickly through the mixed liquor by the beating action of whiplike structures or flagella that enables them to chase after motile dispersed bacteria. This means of capturing bacteria is efficient when the bacterial population is relatively large but inefficient when the bacterial population is relatively small.

Flagellates usually appear in large numbers in the mixed liquor when the bacteria population is growing rapidly or exponentially. With this increase in the bacterial population improvement in treatment efficiency occurs in the mixed liquor as dissolved oxygen concentration increases and pollution decreases. These changes in quantity of substrate and treatment efficiency favor the rapid growth of flagellates, and they appear next as the dominant protozoan group.

FREE-SWIMMING CILIATES

Free-swimming ciliates move through the mixed liquor in a highly controlled fashion by the beating action of short, hairlike structures or cilia. The cilia are found in numerous rows that cover the entire surface of the organism. The cilia not only provide rapid locomotion but also produce water currents that draw bacteria into the mouth opening of the protozoa.

Because the beating action of the cilia is very energy demanding, the protozoa must consume a large quantity of substrate or bacteria. The efficient cropping action of the free-swimming cilia enables these protozoa to out-compete flagellates for dispersed bacteria, and free-swimming ciliates become the dominant protozoan group.

The presence of hardy and hungry free-swimming ciliates and the occurrence of floc formation through sludge aging significantly decrease the number of dispersed bacteria. The number of dispersed bacteria is decreased more through the release

Figure 20.4 Stylonychia. *The crawling ciliate* Stylonchia *has numerous cilia in rows on the ventral surface of the organism and modified cilia or cirri on the posterior portion of the organism. The cirri help anchor* Stylonchia *to the floc particle.*

of secretions from the free-swimming ciliates that coat and removed dispersed bacteria to the developing floc particles.

CRAWLING CILIATES

With increasing treatment efficiency and floc formation, crawling ciliates (Figure 20.4) are next in succession as the dominant protozoan group. Crawling ciliates have cilia only on the ventral surface of the body, where the mouth opening is located. Although the location of the cilia provides for a more efficient feeding mechanism compared to that of the free-swimming ciliates, the reduced number of cilia hinders the ability of the crawling ciliates to control a free-swimming mode of locomotion. Therefore, crawling ciliates prefer to remain on the surface of floc particles.

Some cilia are modified to form "spikes" or cirri (Figure 20.4) that help to anchor the crawling ciliates to the floc particles. Here, the beating of the cilia appears as the movement of many small "legs" as the protozoa "crawl" on the surface of the floc particles. The beating action of the cilia produces water currents that draw dispersed bacteria between the floc particle and the ventral surface of the protozoa to the mouth opening.

With free-swimming ciliates and crawling ciliates feeding upon bacteria, the number of dispersed bacteria decreases in the mixed liquor. This increase in consumption for bacteria coupled with accelerated floc formation results in a significant decrease in the number of dispersed bacteria. The decrease in number of bacteria coupled with improved treatment efficiency affords the opportunity for the proliferation of stalked ciliates. Stalked ciliates have the most efficient feeding mechanism for the capture of dispersed bacteria.

STALKED CILIATES

Stalked ciliates have a band of cilia that surrounds the mouth opening (Figure 20.5). The beating action of the cilia produces water current that draws bacteria directly into the mouth opening. Some stalked ciliates such as *Vorticella* have a contractile filament in the posterior portion or stalk of the organism that permits the organism to "spring." The springing action produces a water vortex that captures bacteria and draws them into the mouth opening.

Although stalked ciliates prefer to remain attached (sessile) to floc particles, they are capable of swimming freely in the bulk solution. Free swimming occurs when

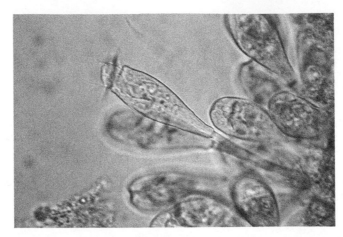

Figure 20.5 Opercularia. *Like all stalked ciliates,* Opercularia *has a band or row of cilia around the mouth opening. The beating action of the cilia produces water currents that draw bacteria into the mouth opening.*

the dissolved oxygen concentration in the mixed liquor drops to <0.5 mg/L. At low dissolved oxygen concentration the stalked ciliate detaches from the floc particle and swims freely in the bulk solution by using the cilia as a "propeller" and the stalk as a "rudder." The stalked ciliate swims in the mixed liquor until it enters a zone of higher dissolved oxygen concentration and then attaches to a floc particle.

Part VI

Rotifers

21

Rotifers

Rotifers (Figure 21.1) or wheel animalcules are aerobic aquatic organisms and are found in numerous habitats including damp soils, sand, and mosses. They enter the activated sludge process through inflow and infiltration (I/I).

Rotifers belong in the phylum Rotifera. The name of the phylum is derived from the ciliated crowns or coronas where the beating action of numerous cilia gives the appearance or optical illusion of rotating wheels. Rotifera, which means "wheel bearers" comes from the Latin words *rota* meaning wheel and *ferre* meaning to bear. The corona consists of two disk-shaped lobes having cilia contracting and expanding at over 1000 beats per minute.

Rotifers are the smallest and simplest of the macroinvertebrates or metazoa. Most rotifers are freshwater organisms, motile and solitary in life. Motility may be free-swimming (Figure 21.2) or crawling (Figure 21.3). Free swimming is achieved by the beating action of the cilia (Figure 21.4). Swimming is slow and forward only, with direction controlled by the foot.

Crawling is achieved through a series of events. In the activated sludge process a crawling rotifer anchors itself to a floc particle by adhesive glands and toes. Next, the rotifer extends its body and attaches its head to the same or a different floc particle. Once the head is anchored, the toes are released from the floc particle, and the body contracts. The contraction of the body places the toes close to the head. Here, the adhesive glands again anchor the toes to the floc particle. The head is then released, and the body is extended. This series of events is repeated over and over as the rotifer moves across floc particles.

Most rotifers are microscopic in size, and the range in lengths of most rotifers is 200–800 μm. There are three basic shapes for most rotifers—sac-shaped, spherical, and worm-shaped (Figure 21.5). A description of rotifers applies to the female.

Microscopic Examination of the Activated Sludge Process, by Michael H. Gerardi
Copyright © 2008 John Wiley & Sons, Inc.

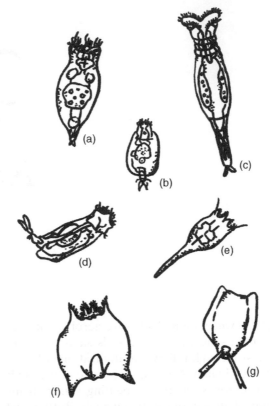

Figure 21.1 *Rotifers. Commonly observed rotifers in the activated sludge process include* Epiphanes (a), Euchlanis (b), Philodina (c), Proales (d), Keratella (e), Platyias (f), *and* Lecane (g).

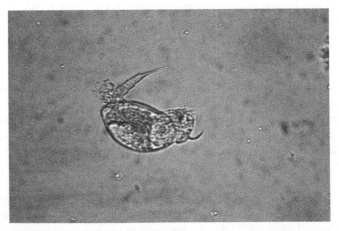

Figure 21.2 Colurella. Colurella *is a free-swimming rotifer.*

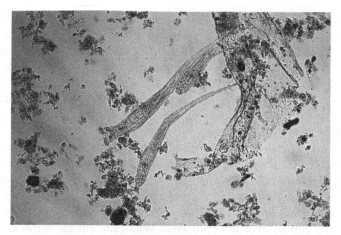

Figure 21.3 Philodina. Philodina *is a crawling rotifer.*

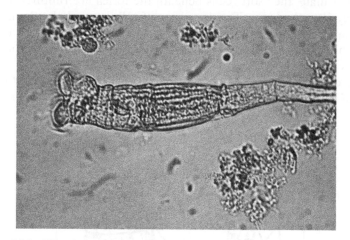

Figure 21.4 *Wheel organ or corona. The head of the rotifer has two bands of cilia.*

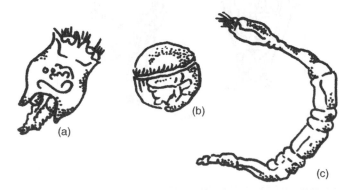

Figure 21.5 *Basic shapes of rotifers. There are three basic shapes for most rotifers. These shapes are sac-shaped* (Noteus) (a), *ball-shaped* (Trochosphaera) (b), *and worm-shaped* (Seison) (c).

Males do not exist in some species and often are sparse in number when they are found. Also, males are much smaller and simpler in structure than females.

The body of the rotifer has three structural zones—the head, the trunk, and the foot (Figure 21.6). The head contains the corona, mouth, and pharynx. The corona forms a funnel, and the pharynx is modified into a muscular mastax or gizzard. The trunk contains most of the organs including the stomach, intestine, cloaca, excretory system, and gonads. The foot is not present in all rotifers, but when present it is tapering in shape and possesses one to four toes.

There are two groups of rotifers as determined by the number of gonads or ovaries present. These groups are the Monogonota (one gonad) and the Bdelloidea (two gonads) (Figure 21.6). Although members of Monogonota have only one gonad, they are more complex in structure than members of Bdelloidea.

An epidermis covers the body of the rotifer. Secretions (scleroproteins) from the epidermis form a cuticle over the epidermis. When the cuticle is thickened it forms a hard protective covering, the lorica. The lorica may have patterns or plates (Figure 21.7). In the presence of cell bursting agents (harsh or slowly degrading surfactants) such as lauryl sulfate, the "soft" cells beneath the lorica are ruptured or dispersed and float away from the lorica. The dispersion of cells leaves behind only the lorica, which often appears as a glowing "horseshoe" as the edge of the lorica reflects the light of the microscope (Figure 21.8). Some rotifers do not have a lorica and are referred to as illoricate rotifers.

Sexes are dimorphic, that is, separate female and male rotifers. Occasionally, copulation of female and male rotifers occurs. However, males are not found in

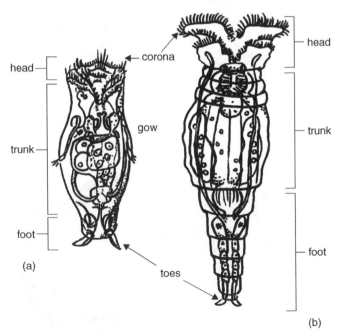

Figure 21.6 *Basic structural zones of rotifers. There are three structural zones for rotifers—head, trunk, and foot. Monogonota rotifers* (a) *have one gonad. Bdelloidea or Digonota rotifers* (b) *have two gonads.*

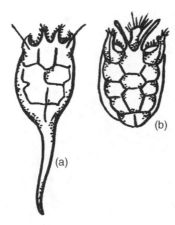

Figure 21.7   *Lorica. The transparent protective covering or lorica is found in many rotifers including Keratella (a). Its lorica (b) consists of a series of plates.*

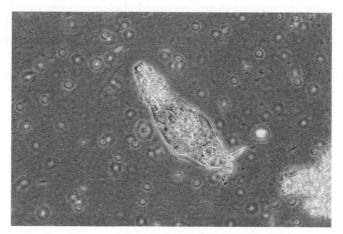

Figure 21.8   *Lorica. The edge of the lorica refracts light from the microscope and appears as a "glowing horseshoe." While the rotifer is dispersed by harsh surfactants, the lorica is not dispersed.*

some species and are infrequently found in some species. When males are not present for reproduction, parthenogenesis occurs with amictic females.

The female rotifer lays eggs (Figure 21.9). Juveniles hatch from the eggs and develop to adults with mature gonads. The maturation of rotifers takes several weeks. Often the MCRT of most activated sludge processes does not permit the maturation of rotifers. Also, rotifers are strict aerobes and are present and active only when dissolved oxygen is at least several mg/L. They are sensitive to adverse operational conditions such as low dissolved oxygen, inhibition, and toxicity. Paralysis and death occur quickly under these operational conditions. Therefore, low MCRT and adverse operational conditions limit the number of rotifers in conventional activated sludge processes. They are present and active in stable operational conditions regardless of MCRT.

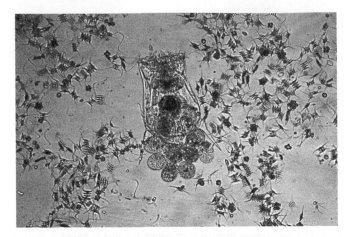

Figure 21.9   *Rotifer eggs.* Brachionus *deposits her eggs.*

Rotifers usually are present in the activated sludge process in relatively small numbers, but they do perform several significant and beneficial roles. They remove colloids, dispersed bacteria, and particulate material through coating and cropping actions. They recycle nutrients, especially nitrogen and phosphorus, in their excreted wastes and serve as bioindicators of the health of the activated sludge process. Rotifers feed on algae, bacteria, detritus, protozoa, and phytoplankton.

Part VII

Worms and Wormlike Organisms

22

Free-Living Nematodes

Free-living nematodes or nemas are terrestrial invertebrates that are associated with a water film in soil and do not cause disease (Figure 22.1). Because they are present in the soil, they enter the activated sludge process through I/I. Nematodes also include eelworms, roundworms, and threadworms.

Most nemas are similar in appearance and are <3 mm in length and <0.05 mm in width. The body consists of three consecutive tubes. The inner tube is the digestive system, the middle tube is the longitudinal muscular system, and the outer tube is the cuticle or "skin." Often, the cuticle is ruptured in the presence of cell busting agents or slowly degrading surfactants (Figure 22.2).

Nemas prefer to be in or on floc particles. Their muscular system provides little control of bodily movement in water, and the nemas display a "whiplike" or thrashing motion in wastewater.

Sexes are separate in nematodes, and reproduction occurs with the deposition of fertilized eggs by the gravid female. These invertebrates are the only significant component of the mixed liquor in large numbers that are capable of burrowing into floc particles, that is, they have a specialized mouth for biting, chewing, crushing, or tearing food or floc particles.

Nemas survive in habitats where dissolved oxygen is high and food is abundant. Their diet consists of algae, bacteria, protozoa, rotifers, other nemas, and detritus. They are present in the mixed liquor in small, often highly variable numbers. They are strict aerobes and do not tolerate low dissolved oxygen concentration and high pollution levels. They are easily observed in the activated sludge process, especially stable processes, regardless of sludge age or MCRT. Because activated sludge processes with high MCRT typically are more stable than those with low MCRT, nematodes usually are found in larger populations in high MCRT processes, but they do not always indicate an old sludge.

Microscopic Examination of the Activated Sludge Process, by Michael H. Gerardi
Copyright © 2008 John Wiley & Sons, Inc.

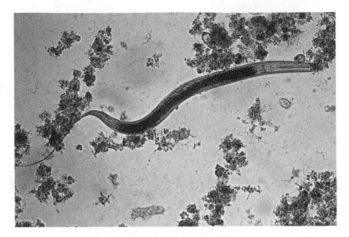

Figure 22.1 *Free-living nematode.*

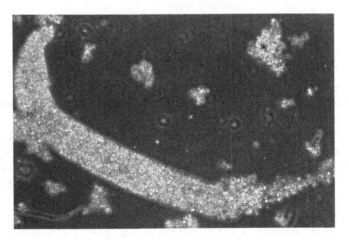

Figure 22.2 *Dispersed cuticle of a free-living nematode.*

Once established in the activated sludge process, free-living nematodes perform many beneficial roles. They promote bacterial and protozoan activity through cropping action. They improve dissolved oxygen, nitrate, nutrient, and substrate penetration into floc particles by their burrowing or tunneling action. They recycle nutrients through their excretions, and their bundles of waste serve as sites for floc formation.

23

Gastrotriches

Gastrotriches (Figure 23.1) are a large group of microscopic (50–3000 μm) animals that are related to rotifers and flatworms and are commonly found in freshwater habitats and enter activated sludge processes through I/I. They are colorless, display bilateral symmetry, and have a complete digestive tract.

Gastrotriches have two terminal projections or adhesive tubes with cement glands that serve in adhesion to vegetation or submerged surfaces. There are two cements glands—one gland secretes the "glue" and one gland secretes the "solvent"

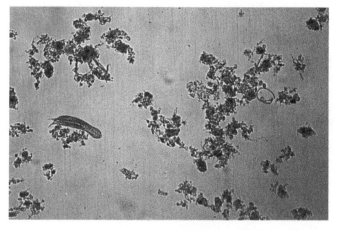

Figure 23.1 *Gastrotrich.*

Microscopic Examination of the Activated Sludge Process, by Michael H. Gerardi
Copyright © 2008 John Wiley & Sons, Inc.

to "stick" and "release," respectively, the adhesive tubes. Gastrotrich feed on bacteria, fungi, protozoa, and dead organic matter by the beating action of four tufts of cilia on the head. They are strict aerobes and do not tolerate low dissolved oxygen concentrations or conditions of organic overload and toxicity. Gastrotriches live only 3–21 days.

24

Water Bears

Water bears or tardigrades (Figure 24.1) are microscopic metazoa (50–1200 μm). They are found in freshwater and damp terrestrial habitats and enter activated sludge processes through I/I.

Water bears consume the fluids of plant and animal cells including amoebae, nematodes, and other water bears. They have a bluntly rounded head with a mouth and eyespots (Figure 24.2). The body is plump and covered with a chitinous cuticle. Water bears have four pairs of appendages or legs that have four to eight claws each. Their body color varies and is caused by pigments in the cuticle, dissolved materials in the body fluid, and content of the digestive tract.

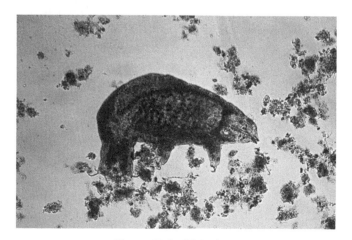

Figure 24.1 *Water bear.*

Microscopic Examination of the Activated Sludge Process, by Michael H. Gerardi
Copyright © 2008 John Wiley & Sons, Inc.

Figure 24.2 Water bear. A lateral view of the water bear shows the organism as having a bluntly rounded head with mouth and eyespot. The body is short, plump, and cylindrical and is covered with a chitinous cuticle. The water bear has four pairs of appendages or legs with claws.

Although water bears are resilient organisms, they are strict aerobes and obtain free molecular oxygen by its movement (diffusion) across the cuticle. Therefore, the absence of dissolved oxygen results in paralysis or death of water bears. Also, conditions such as excess surfactants that affect the integrity of the cuticle and hinder the movement of dissolved oxygen across the cuticle result in paralysis or death of water bears.

25

Bristleworms

Bristleworms or aquatic oligochaetes (Figure 25.1) have the same fundamental structure as the common terrestrial earthworm. Most aquatic oligochaetes are found in the mud and debris substrate of stagnant pools and ponds and in streams and lakes. They are usually <30 mm in length. The body wall is thin, and the internal organs can be easily observed in living specimens.

Bristleworms are segmented, and nearly all segments contain chitinous bristles or setae. The setae may be long, short, straight, curved, sigmoid, or hooked.

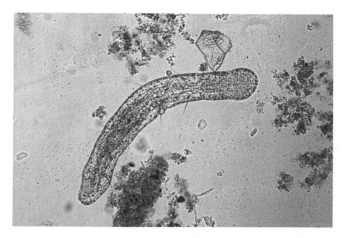

Figure 25.1   *Bristleworm.*

Microscopic Examination of the Activated Sludge Process, by Michael H. Gerardi
Copyright © 2008 John Wiley & Sons, Inc.

Locomotion is a crawling movement similar to that of the earthworm and involves the contraction of the muscular body wall and the use of setae as anchors.

Most aquatic oligochaetes obtain nutriment by ingesting substrate similar to an earthworm. Food usually consists of filamentous algae, diatoms, and animal and plant detritus.

26

Bloodworms

Bloodworms are the larvae of the midge fly or Chironomidae (Class Insecta, Order Diptera). Midge flies are fragile mosquito-like insects that do not bite. Chironomidae go through a complete metamorphosis in their life cycle—egg, larva, pupa, and winged adult midge fly (Figure 26.1). The adult or fly is slender, usually <5 mm in length with long slender wings and legs. They often are mistaken for mosquitoes.

Larvae hatch from eggs. The larvae that contain hemoglobin ("red blood cells") are called bloodworms. Larvae that do not contain hemoglobin are black, brown, green, or transparent. The transparent larvae are known as glassworms.

While skimming the water surface, the female midge fly lays eggs (50–700) in odd-shaped masses. The eggs are deposited on the surface of stagnant or quiet water. Larvae hatch from the eggs in approximately 2 days. Newly hatched larvae are <1 mm in length. Some larvae live freely, while others quickly prepare larval tubes where they reside. The tube is constructed by spinning a loose web of particulate material and silk. During the larval period the larvae molt four times and grow in length to 10–25 mm. At the end of the larval stage, they form pupae that swim to the surface of the water, where they emerge as flying adults.

Midge larvae or bloodworms are often found in wastewater treatment tanks, channels, and ponds. They tolerate well low dissolved oxygen concentrations, low pH values, and high levels of pollution. Bloodworms leave their tubes and are most active at night and when dissolved oxygen concentrations are low. They feed primarily upon algae and organic detritus.

The midge fly can be a severe nuisance. The flies collect in large swarms in late afternoon and evening hours near treatment tanks. They often are attracted to outdoor lights. Swarming flies produce a high-pitched humming sound.

Some control of midge fly and midge fly larvae may be obtained by reducing the amount of algal growth and biofilm on the walls and weirs of treatment tanks

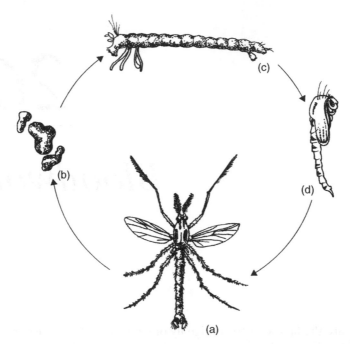

Figure 26.1 *Bloodworm. The midge larva undergoes complete metamorphosis, that is, it has four life stages in its life cycle. These stages are adult* (a), *egg* (b), *larva* (c), *and pupa* (d). *The red larva of the midge fly is known as the bloodworm.*

immediately above and below the water surface. Cleaning the walls or treating the algal growth and biofilm with chlorine may achieve this. Sprinkling water on the surface of the wastewater in clarifiers produces ripples on the surface of the wastewater, helping to prevent egg laying and larval growth. Larvicides may be used if they are approved by the appropriate regulatory agency.

27

Sludge Worms

Sludge worms or *Tubifex* (Figure 27.1) are pollution tolerant and typically proliferate in heavily polluted water. This aquatic worm is also known as an angleworm. They survive in conditions that contain very little dissolved oxygen and can live without dissolved oxygen for short periods of time. In the presence of very low dissolved oxygen concentrations or the absence of dissolved oxygen, sludge worms often may be present in large numbers while other higher life-forms such as rotifers and free-living nematodes are few in number or absent.

Tubifex is a freshwater worm that is capable of living in mud or a semiterrestrial habitat. They enter activated sludge processes through I/I. *Tubifex* is a scavenger.

Figure 27.1 Tubifex *or sludge worm.*

Microscopic Examination of the Activated Sludge Process, by Michael H. Gerardi
Copyright © 2008 John Wiley & Sons, Inc.

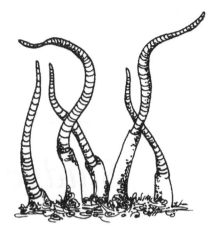

Figure 27.2 Tubifex *head down in tubes.*

Most tubificids live head down in tubes that they build upon detritus, soil, or sludge (Figure 27.2). Here, they consume mostly dead plants and breathe through their epidermis or "skin." The burrowing action of sludge worms in the tube results in aeration of anaerobic sludge.

The sludge worm projects and waves its posterior end from its tube. The waving action circulates water and may facilitate gas exchange, that is, the uptake of oxygen and the release of carbon dioxide.

The segmented, threadlike worm usually is red and typically is more than 25 mm in length. The color of the sludge worm is caused by the presence of dissolved erythrocruorin in the blood. The body of the tubifex consists of a tube within a tube. The outer tube is a soft muscular body wall that is covered with a thin cuticle. The inner tube is the digestive tract that has a terminal mouth and a terminal anus.

Part VIII

Crustaceans

28

Copepods and Cyclops

Copepods (meaning "oar feet") and cyclops are tiny crustaceans. Although they are difficult to see without the use of a microscope or a stereoscopic binocular microscope, the water current that they produce by swimming can be easily observed. They have a quick and jumping motion when swimming. Copepods and cyclops are found in marine and freshwater habitats including wet terrestrial places, under leaf fall, puddles and water-filled recesses of plants, sinkholes, and streambeds. They enter activated sludge processes through I/I. Copepods and cyclops sometimes are found in public water supplies.

Copepods, cyclops, and daphnia are crustaceans that are sometimes found in activated sludge processes. Daphnia belong in the group or order Cladocera. Copepods and cyclops belong in the group or order Copepoda. The term "copepods" is used to describe any free-living subgroup in the Order Copepoda or genus that possesses antennae that are nearly as long as the body of the crustacean (Figure 28.1). Although the term "cyclops" is used to describe any copepods in the subgroup or order Copepoda with antennae that are approximately one-half the length of the body (Figure 28.1), *Cyclops* is a subgroup or genus.

Copepods and cyclops typically are 1–2 mm long and have a teardrop-shaped body and two pairs of antennae. They possess a hard exoskeleton, but the exoske-

Microscopic Examination of the Activated Sludge Process, by Michael H. Gerardi
Copyright © 2008 John Wiley & Sons, Inc.

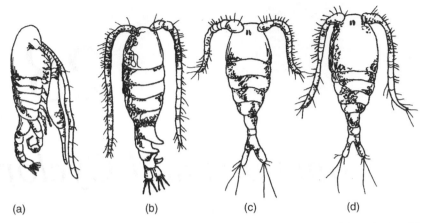

(a)   (b)   (c)   (d)

Figure 28.1   *Copepods and Cyclops. Copepods have antennae that extend from the head to nearly the full length of the body. Copepods include* Diaptomus *(a) and* Epischura *(b).* Cyclops have antenna that extend from the head to approximately half the length of the body. Cyclops include Cyclops *(c) and* Eucyclops *(d).*

leton as well as the entire body is almost transparent. Like all crustaceans, copepods and cyclops molt their exoskeleton in order to grow. They eat bacteria and diatoms, and their presence in large and active numbers is indicative of a relatively nonpolluted environment.

29

Water Fleas

Water fleas or daphnia (Figure 29.1) are small (200–500 μm) crustaceans. They are called water fleas because of their saltatory swimming style or jerky swimming motion. They live in a variety of aquatic habitats, enter activated sludge processes through I/I and feed upon tiny crustaceans and rotifers. Water fleas are often used as bioindicators of biomass health, that is, they are sensitive to changes in wastewater chemistry and their activity and inactivity reflect acceptable and unacceptable operational conditions, respectively.

The body of the water flea is divided into three regions, head, thorax, and abdomen. However, the division between each region is nearly invisible. The thorax contains paired appendages or legs. The beating of the legs produces water current that brings food to the digestive tract. Daphnia feed on algae, bacteria, protozoa, and decaying organic matter.

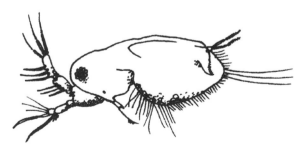

Figure 29.1   *Water flea,* Daphnia.

Microscopic Examination of the Activated Sludge Process, by Michael H. Gerardi
Copyright © 2008 John Wiley & Sons, Inc.

Daphnia are translucent or amber in color. Their body is clam shaped and is covered with a protective chitinous exoskeleton. They have a single eye.

The life span of a water flea is temperature dependent. The average life span is 40–50 days.

30

Ostracoda

Although ostracoda are widely distributed and are occasionally found in activated sludge processes, they have not received the attention that Cladocera, copepoda, and daphnia have received. Ostracoda also are known as "seed shrimp" because without the use of a microscope they look much like small "seeds" with shrimplike appendages (Figure 30.1).

Most freshwater ostracoda are <1 mm in length. The colors of ostracoda include chalky white, black, brown, gray, green, red, and yellow. Ostracoda have two lime-impregnated coverings or valves that provide protection, and together they look like a "seed" under microscopic examination. The valves are connected by an elastic band and a group of muscle fibers. When ostracoda are active, "shrimplike" appendages protrude beneath the valves.

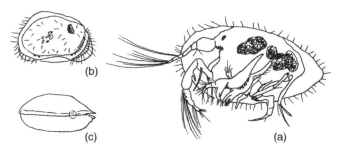

Figure 30.1 *Ostracoda. Several views of ostracoda include a lateral view with valve removed (a); view of ostracoda with the valve over the body of the organisms (b); and a view of ostracoda with both valves covering the organisms (c). With the valves covering the organism, ostracoda resembles a "seed shrimp."*

There is no body segmentation in ostracoda, but the region corresponding to the head has four paired appendages—two pairs of antennae, mandibles, and maxillae. The antennae have short, stiff, clawlike bristles for digging and climbing or long setae for swimming. The mandibles and maxillae are used for feeding purposes. The thoracic region has three pairs of legs. The abdomen has two long caudal appendages or rami with claws. The beating of the antennae and the kicking of rami provide locomotion. Locomotion may appear as creeping to rapid bouncing or scurrying.

Although ostracoda do not occur in grossly polluted waters, they tolerate wide ranges of environmental conditions. They are found on algae, decaying vegetation, and rooted aquatic plants and in gravel, ponds, puddles, and streams. Here, they feed upon algae, bacteria, molds, and fine detritus.

Part IX

Filamentous Organisms

31

Filamentous Organisms

Filamentous organisms (Figure 31.1) or trichomes (rows of bacterial cells that are in close contact with one another) enter activated sludge processes through (1) inflow and infiltration as soil and water organisms, (2) sloughing of the biofilm in the sewer system, and (3) the effluent of biologically pretreated industrial wastewater. Although filamentous algae and filamentous fungi may be observed in activated sludge processes, most filamentous organisms are bacteria. There are approximately 30 filamentous organisms that are found in the activated sludge process. However, 10 of these organisms are responsible for most operational problems associated with the undesired growth of filamentous organisms.

Filamentous organisms perform significant positive and negative roles in the activated sludge process. Positive roles include the degradation of cBOD and the stabilization of floc formation. Filamentous organisms provide a "chain" of strength that enables floc particles to better tolerate turbulence or shearing action and grow in size. However, when filamentous organisms are present in undesired numbers they contribute to settleability problems in the secondary clarifier and loss of solids from the secondary clarifier. In addition to these operational problems caused by filamentous organisms, some filamentous organisms produce foam. The two principal foam-producing filamentous organisms are *Microthrix parvicella* and nocardioforms. The term "thrix" comes from the Greek, meaning hair.

There are specific operational conditions such as high MCRT, low F/M, and low and high pH values that favor the rapid and undesired growth of filamentous organisms (Table 31.1). Therefore, by identifying to name such as *Sphaerotilus natans* or type number such as 0041 the filamentous organisms that are present at undesired numbers, the operational conditions responsible for their proliferation can be identified. Once identified, the activated sludge process can be monitored and corrected

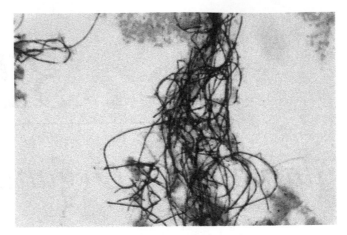

Figure 31.1 *Filamentous organisms. The filamentous organism* Microthrix parvicella *at a relative abundance rating of "5."*

TABLE 31.1 Operational Conditions Associated with the Undesired Growth of Filamentous Organisms

Operational Condition	Filamentous Organism
High MCRT (>10 days)	0041, 0092, 0581, 0675, 0803, 0961, *M. parvicella*, nocardioforms
Fats, oils and grease	0092, *M. parvicella*, nocardioforms
High pH (>7.4)	*M. parvicella*
Low dissolved oxygen and high MCRT	*M. parvicella*
Low dissolved oxygen and low/moderate MCRT	1701, *H. hydrossis*, *S. natans*
Low F/M (<0.05)	021N, 0041, 0092, 0581, 0675, 0803, 0961, *H. hydrossis*, *M. parvicella*, nocardioforms
Low nitrogen or phosphorus	021N, 0041, 0675, 1701, *H. hydrossis*, fungi, nocardioforms, *S. natans*, *Thiothrix* spp.
Low pH (<6.8)	Fungi, nocardioforms
Organic acids	021N, *Beggiatoa* spp., *Thiothrix* spp.
Readily degradable substrates (alcohols, amino acids with sulfur, glucose, volatile fatty acids)	021N, 1851, *H. hydrossis*, nocardioforms, *N. limicola*, *S. natans*, *Thiothrix* spp.
Septicity/sulfides	021N, 0041, *Beggiatoa* spp., *N. limicola*, *Thiothrix* spp.
Slowly degradable substrates	0041, 0092, 0675, *M. parvicella*, nocardioforms
Warm wastewater temperature	1701, *S. natans*
Winter proliferation	*M. parvicella*

for the operational conditions to control the undesired growth of the filamentous organisms.

The identification of the filamentous organisms to name or number is based upon three characteristics and the use of a taxonomic or identification key. These characteristics for identifying filamentous organisms are (1) morphological or structural features, (2) response to specific stains, and (3) response to the sulfur oxidation test or "S" test.

MORPHOLOGICAL FEATURES

The absence or presence of specific structural features that may need to be determined in order to identify filamentous organisms include the following:

- Attached growth (Figure 31.2)
- Branching, false or true (Figures 31.3 and 31.4)
- Cell shape (Figures 31.5 and 31.8)
- Cell size
- Color, translucent or dark
- Constrictions (Figure 31.6)
- Crosswalls or septae (Figure 31.5)
- Filamentous organism location
- Filamentous organism shape (Figures 31.7)
- Filamentous organism size

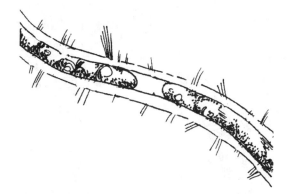

Figure 31.2   *Attached growth. Attached growth on filamentous organism type 0041.*

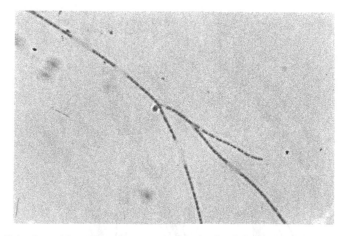

Figure 31.3   *False branching. There is a "gap" or lack of cellular growth between the branches of* Sphaerotilus natans. *Also, there is a transparent sheath that surrounds the branches.*

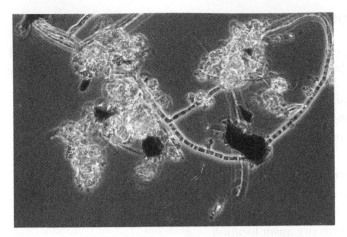

Figure 31.4 *True branching. There is no "gap" between the branches in filamentous fungi. There is continuous cellular growth between the branches, and there is no need for a sheath.*

Figure 31.5 *Cell shapes. There are many shapes of cells in filamentous organisms. Most commonly observed shapes include bacillus, rectangular, square, barrel and disk. Nostocoida limicola has disk-shaped cells. Also, the crosswalls or septae are clearly visible. Crosswalls are the dark lines where two cells adjoin.*

Figure 31.6 *Constrictions. The ends of the bacillus-shaped cells of type 1701 are pinched together like "sausage links."*

Figure 31.7 *Shapes of filamentous organisms. The most commonly occurring shapes for filamentous organisms are curved, coiled, and straight.* Haliscomenobacter hydrossis *is straight or "needlelike."*

Figure 31.8 *Cell shapes. Some filamentous organisms such as type 021N often have cells in more than one shape. Type 021N has barrel-shaped and rectangular cells.*

- Motility
- Sheath (Figure 31.2)
- Sulfur granules, spherical or square (Figure 31.9)

RESPONSE TO SPECIFIC STAINS

Negative and positive responses to specific microbiological stains or staining techniques that may need to be determined in order to identify the filamentous organism include the following:

- Gram
- Neisser
- PHB (poly-β-hydroxybutyrate)
- Sheath

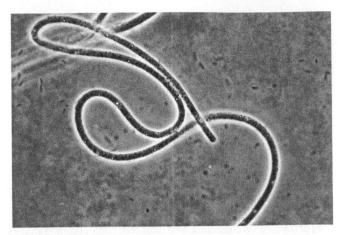

Figure 31.9   Sulfur granules. Highly refractive sulfur granules (clear spots) can be seen along the coiled length of filamentous organism Beggiatoa.

RESPONSE TO THE "S" TEST

The sulfur oxidation test is performed on a mixed liquor sample to determine whether a filamentous organism within the sample is capable of oxidizing sulfur and storing it as granules within the cytoplasm of its individual filamentous cells.

Usually there are two or more filamentous organisms in undesired numbers that contribute to settleability problems and loss of solids. A filamentous organism is present in undesired numbers if its relative abundance rating is "4," "5," or "6" on a scale of "0" to "6," with "0" being "none" and "6" being "excessive" (Table 31.2). Therefore, to ensure that appropriate descriptive features are recorded for each filamentous organism for identification to name or type number a filamentous organism worksheet should be used (Worksheet 31.1). The worksheet should include listings for morphological features, staining responses, and "S" test response.

IDENTIFICATION KEY

On the basis of the morphological features, staining responses, and "S" test response listed in Worksheet 31.1, an appropriate taxonomic or identification key can be used to identify the filamentous organisms (Key 31.1).

Key 31.1   Identification Key for Filamentous Organisms

(*Beggiatoa* spp., flexibacter, fungi, *Haliscomenobacter hydrossis, Microthrix parvicella, Nocardia*/nocardioforms, *Nostocoida limicola, Sphaerotilus natans Thiothrix* spp., Type 0041, Type 0092, Type 0411, Type 0581, Type 0675, Type 0803, Type 094, Type 0961, Type 1701, Type 1851 and Type 021N)

1.   Filamentous organism is motile 2
      Filamentous organism is not motile 3

2.   Filamentous organism is coiled .... *Beggiatoa* spp.
      Filamentous organism is straight .... Flexibacter

TABLE 31.2   Relative Abundance Ratings for Filamentous Organisms

Relative Abundance Rating	Term	Description
0	"None"	Filamentous organisms not observed
1	"Insignificant"	Filamentous organisms present, but found in an occasional floc particle in very few fields of view
2	"Some"	Filamentous organisms present, but found only in some floc particles
3	"Common"	Filamentous organisms observed in most floc particles at low density (1–5 filamentous organisms per floc particle)
4	"Very common"	Filamentous organisms observed in most floc particles at medium density (6–20 filamentous organisms per floc particle)
5	"Abundant"	Filamentous organisms observed in most floc particles at high density (>20 filamentous organisms per floc particle)
6	"Excessive"	Filamentous organisms observed in most floc particles; filamentous organisms more abundant than floc particles; or filamentous organisms growing in large numbers in the bulk solution

WORKSHEET 31.1   Filamentous Organism Identification Worksheet

Feature/Response	Filamentous Organism to be Identified		
	Unknown #1	Unknown #2	Unknown #3
Morphology			
Attached growth			
Branching (yes/no; false/true)			
Cell shape			
Cell size, μm			
Color (transparent or dark)			
Constrictions			
Crosswalls or septae			
Filamentous organism location			
Filamentous organism shape			
Filamentous organism width, μm			
Filamentous organism, length, μm			
Motility			
Sheath			
Sulfur granules (before "S" test)			
Response to stains			
Gram stain (+/−)			
Neisser (+/−)			
PHB (+/−)			
Sheath (+/−)			
Response to "S" test			
"S" test (+/−)			

3. Filamentous organism is branched 4
   Filamentous organism is not branched 5

4. Filamentous organism is falsely branched .... *Sphaerotilus natans*
   Filamentous organism is truly branched and
     Gram-positive .... *Nocardia*
   Filamentous organism is truly branched and
     Gram-negative .... Fungi

5. Filamentous organism is Neisser-positive 6
   Filamentous organism is Neisser-negative 7

6. Filamentous organism is Gram-negative .... Type 0092
   Filamentous organism is Gram-positive
     without a sheath .... *Nostocoida limicola*
   Filamentous organism is Gram-positive, with
     sheath & attached growth .... Type 1851

7. Filamentous organism responds positively to
     the "S" test 8
   Filamentous organism responds negatively to
     the "S" test 9

8. Filamentous organism has basal region thicker
     than apical region .... Type 021N
   Filamentous organisms has uniform thickness
     along its length .... *Thiothrix* spp.

9. Filamentous organism has a sheath 10
   Filamentous organism does not have a sheath 11

10. Filamentous organism is Gram-negative with
      PHB granules .... Type 1701
    Filamentous organism is Gram-negative
      without PHB granules .... *H. hydrossis*
    Thickness of the filamentous organism >1.2 µm .... Type 0041
    Thickness of the filamentous organism <1.2 µm .... Type 0675

11. Filamentous organisms is Neisser-negative
      with positive granules .... *M. parvicella*
    Filamentous organism is Neisser-negative
      without positive granules 12

12. Filamentous organism has PHB granules .... Type 0914
    Filamentous organism does not have PHB
      granules 13

13. Filamentous organism is transparent .... Type 0961
    Filamentous organism is not transparent 14

14.   Filamentous organism is found mostly within
the floc particle .... Type 0581
Filamentous organism is found extending or
free-floating 15

15.   Filamentous organism is straight in shape .... Type 0803
Filamentous organism is bent or irregular in
shape .... Type 0411

Characteristics, growth and control of filamentous organisms

Beggiatoa

Attached growth	Negative	Gram stain	Negative
Branching	Negative	Neisser stain	Negative
Location	Free-floating	PHB stain	Positive
Motility	Positive	"S" test	Positive
Shape	Coiled	Length, μm	100–500
Sheath	Negative	Width, μm	1–3
Growth factors	Organic acids, septicity, sulfides		
Control measures	Control of growth factors, use of oxidizing agent (chlorine, hydrogen peroxide)		

Haliscomenobacter hydrossis

Attached growth	Negative/positive	Gram stain	Negative
Branching	Negative	Neisser stain	Negative
Location	Extending, free-floating	PHB stain	Negative
Motility	Negative	"S" test	Negative
Shape	Straight	Length, μm	20–100
Sheath	Positive	Width, μm	0.5
Growth factors	Low dissolved oxygen, low F/M, low nitrogen or phosphorus, readily degradable substrates,		
Control measures	Control of growth factors, use of oxidizing agent (chlorine or hydrogen peroxide), increase SRT, use of aerobic, anoxic, or anaerobic selector		

Microthrix parvicella

Attached growth	Negative	Gram stain	Positive
Branching	Negative	Neisser stain	Negative
Location	Inside	PHB stain	Positive
Motility	Negative	"S" test	Negative
Shape	Coiled	Length, μm	100–400
Sheath	Negative	Width, μm	0.8
Growth factors	High MCRT, fats, oils, and grease, high pH, low dissolved oxygen, low F/M, slowly degradable substrates, winter proliferation		
Control measures	Control of growth factors, use of oxidizing agent (chlorine or hydrogen peroxide), maintain uniform DO level in aeration tank		

Nocardia/nocardioforms

Attached growth	Negative	Gram stain	Positive
Branching	Positive	Neisser stain	Negative
Location	Inside	PHB stain	Positive
Motility	Negative	"S" test	Negative
Shape	Irregular	Length, μm	10–20
Sheath	Negative	Width, μm	1
Growth factors	Fats, oils and grease, low F/M, low pH, readily degradable substrates, slowly degradable substrates		
Control measures	Control of growth factors, use of oxidizing agent (chlorine or hydrogen peroxide), use of anoxic selector		

Nostocoida limicola

Attached growth	Negative	Gram stain	Positive
Branching	Negative	Neisser stain	Positive
Location	Inside, extending	PHB stain	Positive/negative
Motility	Negative	"S" test	Negative
Shape	Coiled	Length, μm	100–300
Sheath	Negative	Width, μm	1.2–1.4
Growth factors	Readily degradable substrates, septicity, sulfides		
Control measures	Control of growth factors, use of oxidizing agent (chlorine or hydrogen peroxide), use of aerobic, anoxic, or anaerobic selector		

Sphaerotilus natans

Attached growth	Negative	Gram stain	Negative
Branching	Positive	Neisser stain	Negative
Location	Extending	PHB stain	Positive
Motility	Negative	"S" test	Negative
Shape	Curved, straight	Length, μm	>500
Sheath	Positive	Width, μm	1.0–1.4
Growth factors	Low dissolved oxygen, low nitrogen or phosphorus, warm wastewater temperature		
Control measures	Control of growth factors, use of oxidizing agent (chlorine or hydrogen peroxide), increase SRT, use of aerobic, anoxic, or anaerobic selector		

Thiothrix

Attached growth	Negative	Gram stain	Negative
Branching	Negative	Neisser stain	Negative
Location	Extending	PHB stain	Positive
Motility	Negative	"S" test	Positive
Shape	Straight, curved	Length, μm	50–200
Sheath	Positive	Width, μm	0.8–1.4
Growth factors	Low nitrogen or phosphorus, readily degradable substrates, septicity, sulfides		
Control measures	Control of growth factors, use of oxidizing agent (chlorine or hydrogen peroxide), use of aerobic, anoxic, or anaerobic selector		

Type 0041

Attached growth	Positive	Gram stain	Positive
Branching	Negative	Neisser stain	Negative
Location	Inside, extending	PHB stain	Negative
Motility	Negative	"S" test	Negative
Shape	Straight	Length, μm	100–500
Sheath	Positive	Width, μm	1.4–1.6
Growth factors	High MCRT, low F/M, low nitrogen or phosphorus, septicity, sulfides, slowly degradable substrate		
Control measures	Control of growth factors, use of oxidizing agent (chlorine or hydrogen peroxide), maintain uniform DO level in aeration tank		

Type 0092

Attached growth	Negative	Gram stain	Negative
Branching	Negative	Neisser stain	Positive
Location	Inside	PHB stain	Negative
Motility	Negative	"S" test	Negative
Shape	Straight, bent	Length, μm	20–60
Sheath	Negative	Width, μm	0.8–1
Growth factors	High MCRT, fats, oils, and grease, low F/M, slowly degradable substrates		
Control measures	Control growth factors, use of oxidizing agent (chlorine or hydrogen peroxide), maintain uniform DO in aeration tank		

Type 0581

Attached growth	Negative	Gram stain	Negative
Branching	Negative	Neisser stain	Negative
Location	Inside	PHB stain	Negative
Motility	Negative	"S" test	Negative
Shape	Coiled	Length, μm	100–200
Sheath	Negative	Width, μm	0.5–0.8
Growth factors	High MCRT, low F/M		
Control measures	Control of growth factors, use of oxidizing agent (chlorine or hydrogen peroxide)		

Type 0675

Attached growth	Positive	Gram stain	Positive
Branching	Negative	Neisser stain	Negative
Location	Inside	PHB stain	Negative
Motility	Negative	"S" test	Negative
Shape	Straight	Length, μm	50–150
Sheath	Positive	Width, μm	0.8–1
Growth factors	High MCRT, low F/M, low nitrogen or phosphorus, slowly degradable substrates		
Control measures	Control of growth factors, use of oxidizing agent (chlorine or hydrogen peroxide, maintain uniform DO in aeration tank		

Type 0803

Attached growth	Negative	Gram stain	Negative
Branching	Negative	Neisser stain	Negative
Location	Extending	PHB stain	Negative
Motility	Negative	"S" test	Negative
Shape	Straight	Length, μm	50–150
Sheath	Negative	Width, μm	0.8
Growth factors	High MCRT, low F/M		
Control measures	Control of growth factors, use of oxidizing agent (chlorine or hydrogen peroxide)		

Type 0961

Attached growth	Negative	Gram stain	Negative
Branching	Negative	Neisser stain	Negative
Location	Extending	PHB stain	Negative
Motility	Negative	"S" test	Negative
Shape	Straight	Length, μm	40–80
Sheath	Negative	Width, μm	0.8–1.2
Growth factors	High MCRT		
Control measures	Control of growth factors, use of oxidizing agent (chlorine or hydrogen peroxide)		

Type 1701

Attached growth	Positive	Gram stain	Negative
Branching	Negative	Neisser stain	Negative
Location	Inside, extending	PHB stain	Positive
Motility	Negative	"S" test	Negative
Shape	Straight, bent	Length, μm	20–80
Sheath	Positive	Width, μm	0.6–0.8
Growth factors	Low dissolved oxygen, low nitrogen or phosphorus, warm wastewater temperature		
Control measures	Control of growth factors, use of oxidizing agent (chlorine or hydrogen peroxide), increase SRT, use of aerobic, anoxic, or anaerobic selector		

Type 1851

Attached growth	Negative,/positive	Gram stain	Positive
Branching	Negative	Neisser stain	Negative
Location	Extending	PHB stain	Negative
Motility	Negative	"S" test	Negative
Shape	Straight, bent	Length, μm	100–300
Sheath	Positive	Width, μm	0.8
Growth factors	Readily degradable substrates		
Control measures	Control of growth factors, use of oxidizing agent (chlorine or hydrogen peroxide)		

Type 021N

Attached growth	Negative	Gram stain	Negative
Branching	Negative	Neisser stain	Negative
Location	Extending	PHB stain	Positive
Motility	Negative	"S" test	Positive
Shape	Straight, coiled	Length, μm	50 to >500
Sheath	Negative	Width, μm	1–2
Growth factors	Low F/M, organic acids, readily degradable substrates, septicity, sulfides		
Control measures	Control of growth factors, use of oxidizing agent (chlorine or hydrogen peroxide), use of aerobic, anoxic, or anaerobic selector		

Part X

Algae and Fungi

32

Algae

Blue-green and green algae (Figures 32.1 and 32.2) are observed infrequently in the mixed liquor because of the lack of penetration of sunlight and the presence of a turbulent environment produced through aeration and mixing patterns. However, algae grow on the weirs and walls of secondary clarifiers and are found in the effluents of roughing towers (trickling filters) ahead of aeration tanks and industrial pretreatment facilities using fixed film media (trickling filters and rotating biological contactors). Therefore, the return of activated sludge (RAS) and the effluents from

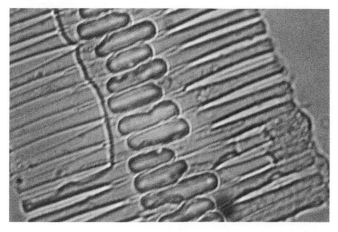

Figure 32.1 Flagellaria.

Microscopic Examination of the Activated Sludge Process, by Michael H. Gerardi
Copyright © 2008 John Wiley & Sons, Inc.

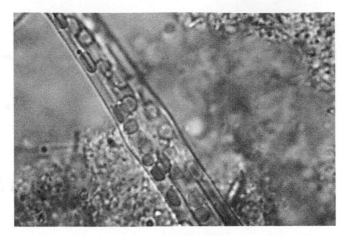

Figure 32.2   Tabellaria.

fixed film processes seed the mixed liquor with algae. Commonly observed algae are unicellular such as *Chlorella* and filamentous such as *Anabaena, Oscillatoria,* and *Spirogyra. Chlorella* and *Spirogyra* are green algae, while *Anabaena* and *Oscillatoria* are blue-green algae.

33

Fungi

There are three fungal groups of concern in the activated sludge process. These groups are (1) the pathogenic fungi, (2) the unicellular fungi, and (3) the filamentous fungi.

There are approximately 50 pathogenic fungi. With respect to pathogenic fungi in wastewater treatment facilities that present risk of infection to wastewater personnel, there are two, *Candida* and *Aspergillus fumigatis*. *Candida*, which causes oral and vaginal infections known as candidiasis, is a potential risk to all wastewater personnel. However, *Aspergillus fumigatis* (Figure 33.1), which causes the respiratory tract infection known as aspergillosis, is an actual risk to operators of thermophilic composting facilities.

Unicellular fungi are single-celled organisms and are often represented by the yeast *Saccharomyces*. Unicellular fungi are widely distributed in nature. They are associated with other microorganisms as part of the normal inhabitants of soil, vegetation, and aquatic environments. They enter activated sludge processes through I/I; industrial discharges including alcoholic beverages, baking, bioremediation, industrial ethanol production, nutritional supplements, yeast extract, and food spoilage; and fermentative conditions in sewer systems.

Unicellular fungi are capable of degrading a large variety of substrates or cBOD. Many of these substrates cannot be degraded or are slowly degraded by bacteria. Therefore, unicellular fungi contribute to increased treatment efficiency of wastewater and often are added to bioaugmentation products (liquid and dry cultures of commercially prepared bacteria) to improve bioaugmentation performance.

Unfortunately, filamentous fungi (Figure 33.2) contribute to settleability problems in the activated sludge process. Filamentous fungi are truly branched organisms. Although the Gram stain is used to differentiate Gram-negative (red) and

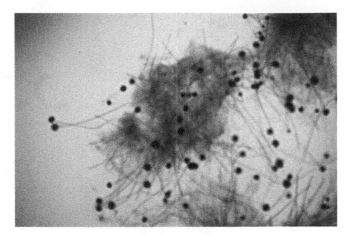

Figure 33.1 Aspergillus fumigatis. Aspergillus fumigatis *has a vegetative state (filamentous) and a spore state (black dots).*

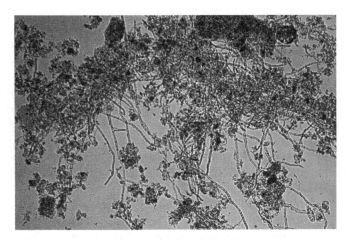

Figure 33.2 *Filamentous fungi.*

Gram-positive (blue) bacteria because of their differences in cell wall composition, filamentous fungi or the cell wall of the filamentous fungi stain red when exposed to Gram staining.

Filamentous fungi grow in the presence of sugars, organic acids, and other easily metabolized carbon sources. Their rapid growth or proliferation occurs at low pH values and nutrient deficiencies, especially for severe phosphorus deficiency. Filamentous fungi can grow at pH values <6. Because fungi require less nitrogen for growth than bacteria, fungi obtain a competitive edge over bacteria during low nitrogen levels and quickly proliferate. In activated sludge processes that receive relatively large quantities of antibiotic wastes, filamentous fungi may also obtain a competitive edge over bacteria.

Part XI

Collection, Evaluation, and Presentation of Observations

34

Microscopic Set-up and Rating Tables

Although numerous wastewater samples (Table 34.1) can be used to evaluate the health of the treatment process, the mixed liquor is typically used. Major or significant components of microscopic examinations that are evaluated vary from treatment plant to treatment plant but often include bulk solution, floc particles, filamentous organisms, protozoa, and metazoa (Table 34.2). If an undesired quantity of foam is present on the mixed liquor, microscopic examination of foam also may be performed (Table 34.3).

BULK SOLUTION

The bulk solution is the water that surrounds the floc particles. In healthy mixed liquor, the bulk solution contains little or an insignificant amount of dispersed growth or particulate material. Dispersed growth and particulate material are removed from the bulk solution through three major mechanisms. First, if the charge on the dispersed growth or particulate material is compatible for adsorption to the surface of the floc particles, they are quickly adsorbed. Second, if the charge is not compatible for adsorption, the charge may be made compatible through the coating action of secretions released by ciliated protozoa and metazoa. Third, dispersed growth may be cropped or consumed by ciliated protozoa and metazoa.

When floc formation is interrupted by an adverse operational condition, dispersed growth and particulate material may not be removed from the bulk solution and adsorbed to the floc particles and may be released by floc particles. Interruption of floc formation often results in a significant increase in dispersed growth and particulate material in the bulk solution. There are numerous operational conditions

TABLE 34.1   Wastewater Samples Available for Microscopic Examination

Centrate or filtrate from sludge dewatering operations
Digester (aerobic or anaerobic) decant
Digester (aerobic or anaerobic) solids
Final effluent
Foam
Industrial effluent from biological pretreatment systems
Leachate
Mixed liquor
Return activated sludge
Scum
Secondary clarifier effluent
Settleability test, bubbles/foam
Settleability test, floating solids
Settleability test, settled solids
Settleability test, supernatant
Thickener overflow

TABLE 34.2   Significant Components of Microscopic Examinations of Mixed Liquor

Component	Feature
Bulk solution	Dispersed growth, relative abundance
	Particulate material, relative abundance
Floc particles	Shape, dominant
	Size, dominant
	Size, range
	Strength, weak or firm
	Strength, weak perimeter and firm core
	India ink reverse stain, negative or positive
Zoogloeal growth	Relative abundance
Floc particle/filament composition	Interfloc bridging, insignificant or significant
	Open floc formation, insignificant or significant
Filamentous organisms	Length or range in lengths of most filaments
	Location of most filaments
	Relative abundance rating of all filaments
	Dominant filaments
	Recessive filaments
Protozoa	Activity
	Structure
	Count or relative abundance
	Profile or percent composition of protozoan groups
	Dominant protozoa
Metazoa (rotifers and nematodes)	Activity
	Structure
	Count or relative abundance

TABLE 34.3   Significant Components of Microscopic Examinations of Foam

Foam-producing filamentous organisms
Nutrient-deficient floc particles
Zoogloeal growth

TABLE 34.4   Operational Conditions Responsible for the Interruption of Floc Formation

Operational Condition	Description or Example
Anionic detergent or cell busting agent	Lauryl sulfate
Colloidal floc	Excess proteinaceous wastes
Elevated temperature	>32 °C
Foam production	Foam-producing filamentous organisms
High pH/low pH	>8.5/<6.5
Increase in MLVSS	Accumulation of fats, oils, and grease
Lack of ciliated protozoa	<100 per milliliter
Low dissolved oxygen concentration	<1 mg/L for 10 consecutive hours
Nutrient deficiency	Usually nitrogen or phosphorus
Salinity	Excess manganese, sodium, and/or potassium
Scum production	Die-off of large numbers of bacteria
Septicity	<−100 millivolts (ORP)
Shearing action	RAS pump, surface aeration
Slug discharge of soluble cBOD	3× normal loading of soluble cBOD
Sulfates	>500 mg/L
Total dissolved solids (TDS)	>5000 mg/L
Toxicity	Excess chlorination of RAS
Undesired filamentous organism growth	Relative abundance rating >"3"
Viscous floc or zoogloeal growth	Rapid floc-forming bacterial growth
Young sludge age	<3 days MCRT

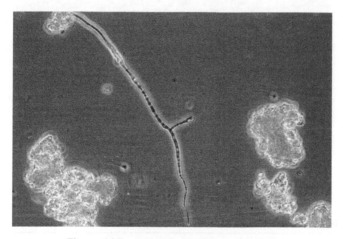

Figure 34.1   Insignificant dispersed growth.

that are responsible for the interruption of floc formation. These conditions are listed in Table 34.4.

BULK SOLUTION, DISPERSED GROWTH

Dispersed growth consists of very small floc particles (<10 μm) that appear as small "dots" when examined under the microscope at 100× total magnification. The amount of dispersed growth can be rated as "insignificant," "significant", or "excessive" (Figures 34.1, 34.2, and 34.3, respectively) (Table 34.5). To observe

Figure 34.2 Significant dispersed growth.

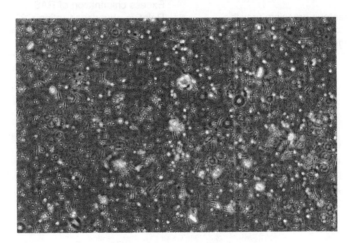

Figure 34.3 Excessive dispersed growth.

TABLE 34.5   Relative Abundance Ratings for Dispersed Growth

Rating	Description
Insignificant	<20 "dots" per field of view
Significant	Tens of "dots" per field of view, for example, 20, 30, 40 . . .
Excessive	Hundreds of "dots" per field of view, for example, 100, 200, 300 . . .

dispersed growth a wet mount of mixed liquor can be examined under the bright-field or phase contrast microscope (Table 34.6). A drop of methylene blue may be added to the wet mount to more easily observe the dispersed growth when using the bright-field microscope. It is not necessary to count each "dot" in each field of view as the wet mount is scanned. It is only necessary to subjectively evaluate the relative amount of dispersed growth.

TABLE 34.6 Microscopic Set-up for Evaluating Dispersed Growth

Slide Preparation	Microscope	Power of Magnification
Wet mount	Bright-field or phase contrast	100× total

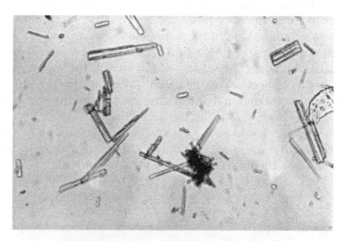

Figure 34.4 *Particulate material, plastic resins.*

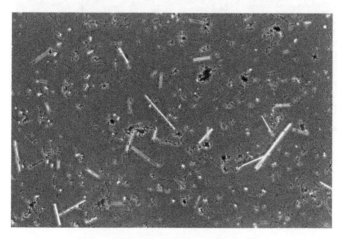

Figure 34.5 *Particulate material, fibers.*

Healthy mixed liquor should have a relative abundance rating of "insignificant" for dispersed growth. Ratings of "significant" and "excessive" indicate the need for identification and correction of a potentially adverse operational condition.

BULK SOLUTION, PARTICULATE MATERIAL

Particulate material consists of nonliving or inert particles. Particulate material may be relative small, for example, plastic resins (<20 μm) (Figure 34.4) or relatively large, for example, fibrous material (>1000 μm) (Figure 34.5). Particulate material may be any color, texture, or shape and should be found adsorbed to the surface of floc particles or within floc particles (Figure 34.6). Particulate material is considered to be inert particles that are >10 µm in size.

The amount of particulate material can be rated as "insignificant" or "significant" (Table 34.7). To observe particulate material a wet mount of mixed liquor can be examined under the bright-field or phase contrast microscope (Table 34.8). A drop of methylene blue may be added to the wet mount to more easily observe the particulate material when using the bright-field microscope. It is not necessary to count each piece of particulate material in each field of view as the wet mount is scanned. It is only necessary to subjectively evaluate the amount of particulate material that is adsorbed to or incorporated in floc particles as compared to the amount of particulate material that is free-floating in the bulk solution.

Healthy mixed liquor should have a relative abundance rating of "insignificant" for particulate material. A rating of "significant" indicates the need for identification and correction of a potentially adverse operational condition.

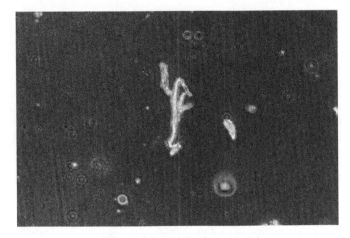

Figure 34.6 *Particulate material.*

TABLE 34.7   Relative Abundance Ratings for Particulate Material

Rating	Description
Insignificant	Most particulate material adsorbed to or incorporated in floc particles
Significant	Most particulate material free-floating in the bulk solution

TABLE 34.8   Microscopic Set-up for Evaluating Particulate Material

Slide Preparation	Microscope	Power of Magnification
Wet mount	Bright-field or phase contrast	100× total

FLOC PARTICLES, DOMINANT SHAPE

There are two frequently observed shapes in mixed liquor and one infrequently observed shape. The frequently observed shapes are spherical (Figure 34.7) and irregular (Figure 34.8), while the infrequently observed shape is oval (Figure 34.9).

Typically, the spherical and irregular shapes can be found together in nearly all activated sludge processes. In a mature activated sludge process with significant growth of filamentous organisms, the irregular shape is dominant. The filamentous organisms provide a network of strength that resists shearing action or turbulence in the treatment process. Therefore, the floc bacteria grow and agglutinate or flocculate along the lengths of the filamentous organisms as the floc particle increases in size. In a young activated sludge process very few filamentous organisms are present, and resistance to shearing action is poor. Therefore, floc bacteria simply

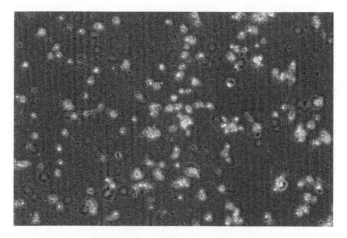

Figure 34.7 *Spherical floc particles.*

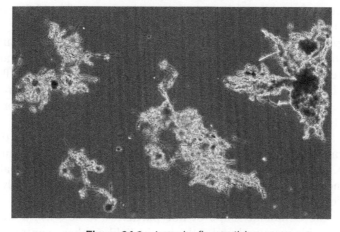

Figure 34.8 *Irregular floc particles.*

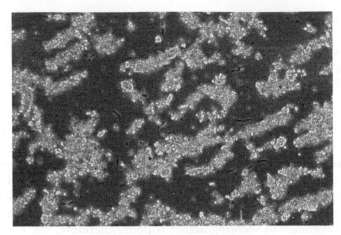

Figure 34.9   *Oval floc particles.*

TABLE 34.9   Floc Particle Sizes and Shapes

Size	Shape	Comment
Small	Spherical	Typical of young sludge
	Irregular	Shearing action possible
Medium	Irregular	Typical of old sludge
	Spherical	Surfactant discharge, possible
	Oval	Metals discharge, possible
Large	Irregular	Typical of old sludge
	Spherical	Surfactant discharge, possible
	Oval	Metals discharge, possible

flocculate in small spherical particles until filamentous organisms grow in significant numbers.

Because young bacteria produce a relatively small quantity of oils that are adsorbed to the floc particle, young floc particles are white in color. However, old bacteria produce a relatively large quantity of oils that are adsorbed to the floc particle. Because of the adsorption of a large quantity of oils, old floc particles are golden-brown in color.

Oval floc particles are infrequently observed and usually are a result of a dirty microscope slide or an undesired operational condition. Dirty slides and oily film on new or "precleaned" microscope slides permit the formation of oval floc particles. Therefore, it is important that all microscope slides be cleaned before use. Slides may be cleaned with Ajax™ or a similar compound and then rinsed thoroughly with deionized water.

Oval floc particles may be present in oily wastewater or wastewater that contains excess multivalent cations including metals from industrial discharges and coagulant such as alum, ferric, or lime. Oval floc particles produced by multivalent cations indicate the need for identification and correction of a potentially adverse operational condition.

To observe the dominant shape of floc particles (Table 34.9) a wet mount of mixed liquor can be examined under the bright-field or phase contrast microscope

TABLE 34.10   Microscopic Set-up for Evaluating Shapes of Floc Particles

Slide Preparation	Microscope	Power of Magnification
Wet mount	Bright-field or phase contrast	100× total

TABLE 34.11   Commonly Used Size Ranges for Floc Particles

Small	Medium	Large
<150 μm	150–500 μm	>500 μm

TABLE 34.12   Microscopic Set-up for Evaluating Sizes of Floc Particles

Slide Preparation	Microscope	Power of Magnification
Wet mount	Bright-field or phase contrast	100× total

(Table 34.10). A drop of methylene blue may be added to the wet mount to more easily observe the shapes of the floc particles when using the bright-field microscope. It is not necessary to record the shape of each floc particle in each field of view as the wet mount is scanned. It is only necessary to subjectively evaluate the relative abundance of each shape.

FLOC PARTICLES, DOMINANT SIZE

Floc particles commonly are placed into one of three size groups (Table 34.11), and the size of the floc particle is measured in microns (μm) with an ocular micrometer. The three groups of floc particles according to size are small (<150 μm), medium (150–500 μm), and large (>500 μm). In a healthy activated sludge process most floc particles are medium and/or large in size. To observe the dominant size of floc particles a wet mount of mixed liquor can be examined under the bright-field or phase contrast microscope (Table 34.12). A drop of methylene blue may be added to the wet mount to more easily observe the shapes of the floc particles when using the bright-field microscope.

FLOC PARTICLES, RANGE IN SIZES

Usually one or two sizes of floc particles dominate an activated sludge process. Changes in dominant sizes may be indicative of significant changes in operational conditions, for example, increasing sludge age or excess wasting of solids. Although one or two sizes of floc particles are dominant in an activated sludge process, the range in sizes, for example, from 80 to 1900 μm to 20 to 1800 μm, also may be indicative of significant changes in operational conditions such as the rapid young growth from a slug discharge of soluble cBOD.

To observe the dominant size of floc particles a wet mount of mixed liquor can be examined under the bright-field or phase contrast microscope (Table 34.13). A

TABLE 34.13   Microscopic Set-up for Evaluating Range in Sizes of Floc Particles

Slide Preparation	Microscope	Power of Magnification
Wet mount	Bright-field or phase contrast	100× total

TABLE 34.14   Characteristics of Firm and Weak Floc Particles under Methylene Blue Stain

Floc Particle Strength	Characteristics
Firm	Dark blue particle with few openings or clear areas
Weak	Light blue particle with several openings or clear areas

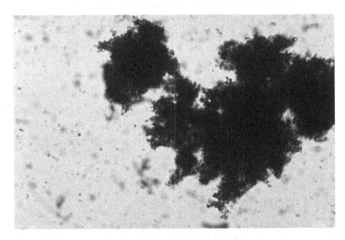

Figure 34.10   *Firm floc particle under methylene blue. The majority of the area of the floc particle is dark blue.*

drop of methylene blue may be added to the wet mount to more easily observe the shapes of the floc particles when using the bright-field microscope.

FLOC PARTICLES, STRENGTH

The strength of floc particles is an important characteristic. If floc particles are firm, the particles are resistant to shearing action. If floc particles are weak, the particles may be easily sheared in the activated sludge process.

Determination of the strength of the floc particles is a subjective evaluation of the compaction of the floc bacteria (Table 34.14). The bacteria may be tightly adjoined (firm), or the floc bacteria may be loosely adjoined (weak). The relative strength of the floc particles may be viewed through methylene blue staining.

Firm floc particles appear dark blue throughout the majority of the area of the floc particle. (Figure 34.10) Firm floc particles also possess very few, if any, openings or clear areas under methylene blue stain. Weak floc particles appear light blue throughout the majority of the area of the floc particle (Figure 34.11). Weak floc particles also possess openings or clear areas under methylene blue stain.

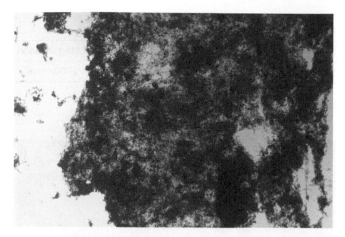

Figure 34.11 *Weak floc particle under methylene blue. Much of the area of the floc particle is light blue, and openings or voids appear in the floc particle.*

TABLE 34.15 Microscopic Set-up for Evaluating the Relative Strength of Floc Particles

Slide Preparation	Microscope	Power of Magnification
Wet mount	Bright-field	100× total

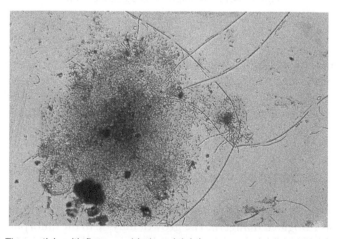

Figure 34.12 *Floc particle with firm core (dark and tightly compacted cells) and weak perimeter (light and loosely compacted cells) under the safranin stain.*

To observe the relative strength of floc particles, a wet mount of mixed liquor can be examined under the bright-field microscope (Table 34.15). A drop of methylene blue must be added to the wet mount to observe the compaction of bacterial cells.

Floc particles that have experienced a slug discharge of soluble cBOD grow very rapidly. This rapid growth occurs mostly on the perimeter of the floc particles and results in the production of a weak or loosely flocculated mass of bacteria on the perimeter of the floc particle (Figure 34.12). However, the core of the floc particle

TABLE 34.16   Microscopic Set-up for Observing Weak Bacterial Growth from a Slug Discharge of Soluble cBOD

Slide Preparation	Microscope	Power of Magnification
Smear	Bright-field	400× or 1000× total

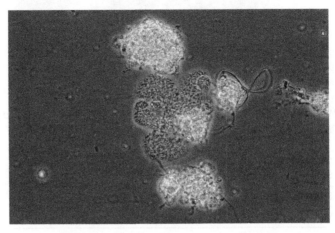

Figure 34.13   *Amorphous zoogloeal growth.*

contains a firm or tightly flocculated mass of old bacteria. The difference in compaction and age of the floc bacteria can be observed with a safranin stain. To observe the rapid and weak growth of bacteria from a slug discharge of soluble cBOD, a safranin smear of mixed liquor can be examined under the bright-field microscope (Table 34.16).

ZOOGLOEAL GROWTH OR VISCOUS FLOC

Zoogloeal growth or viscous floc is the rapid and undesired growth of floc-forming bacteria such as *Zoogloea ramigera*. The growth may be in the amorphous (globular) form (Figure 34.13) or the dendritic (fingerlike) form (Figure 34.14). Usually one form of the growth is present or dominant.

Zoogloeal growth results in the production of weak and buoyant floc particles. The growth may be associated with billowy white foam, and zoogloeal growth may occur of the side walls and weirs of secondary clarifiers.

Zoogloeal growth may be rated as "insignificant' or "significant" (Table 34.17) Insignificant indicates that few fields of view contain zoogloeal growth. Significant indicates that most fields of view contain zoogloeal growth. Zoogloeal growth may be observed with the bright-field or phase contrast microscope (Table 34.18). However, the use of methylene blue may enhance the appearance of zoogloeal growth.

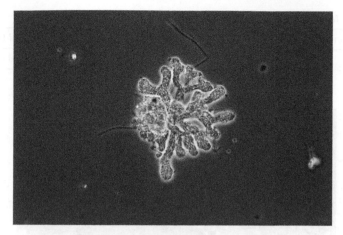

Figure 34.14 *Dendritic zoogloeal growth.*

TABLE 34.17   Table Ratings for Zoogloeal Growth

Rating	Description
Insignificant	Few fields of view contain zoogloeal growth
Significant	Most fields of view contain zoogloeal growth

TABLE 34.18   Microscopic Set-up for Evaluating Zoogloeal Growth

Slide Preparation	Microscope	Power of Magnification
Wet mount	Bright-field or phase contrast	100× total

NUTRIENT DEFICIENCY (INDIA INK REVERSE STAIN)

The india ink reverse stain is used to determine the probability of a nutrient deficiency (usually nitrogen or phosphorus) based on the relative amount of stored food within the floc particles. With increasing amounts of stored food there is a increasing probability that the floc particle is nutrient deficient, that is, the treatment process has experienced a nutrient deficiency. During the nutrient deficiency soluble cBOD that could not be degraded was converted to insoluble polysaccharides and stored within the floc particles. When nutrients become available, the insoluble polysaccharides are then solubilized and degraded.

The staining technique for the india ink reverse stain is performed with an aqueous solution of India ink (china ink or nigrosine) and a phase contrast microscope (Table 34.19). During the staining technique, the carbon black particles within the ink penetrate the floc particles. The areas where carbon black particles are found appear black or golden-brown under phase contrast microscopy (Figure 34.15). The areas where carbon black particles are absent appear white (Figure 34.16). The absence of the carbon black particles is caused by the presence of stored food that prevents the movement of carbon black particles.

By subjectively evaluating the relative amount of the area of the floc particle that is black and/or golden-brown as compared to the amount of the area of the floc

TABLE 34.19 Microscopic Set-up for the India Ink Reverse Stain

Slide Preparation	Microscope	Power of Magnification
Wet mount	Phase contrast	100× total or 1000× total

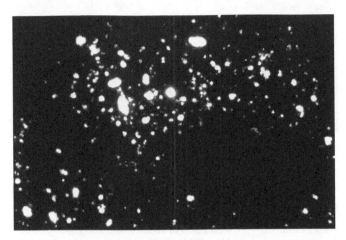

Figure 34.15 *Negative india ink reverse stain.*

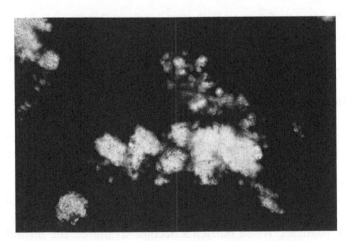

Figure 34.16 *Positive india ink reverse stain.*

particle that is white, the probability of a nutrient deficiency can be determined (Table 34.20). If the majority of the area of the floc particle is black and/or golden-brown, the probability of a nutrient deficiency is relatively low. However, if the majority of the area of the floc particle is white, the probability of a nutrient deficiency is relatively high.

False positives may occur during an India ink reverse stain if large amounts of gelatinous materials are present. These materials, like stored food, hinder the movement of carbon black into the floc particles. Gelatinous materials may be produced

TABLE 34.20   Ratings for the India Ink Reverse Stain

Rating	Description
Negative	Area of most floc particles is black and/or golden-brown; only a small amount of white or "spotting" of white is present
Positive	Area of most floc particles is white

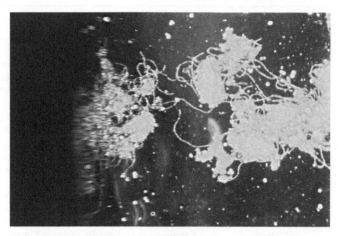

Figure 34.17 *Poor preparation of the india ink stain. Inadequate amount of india ink added to the mixed liquor.*

during zoogloeal growth and toxicity. Zoogloeal growth may be confirmed by examining the floc particles for amorphous or dendritic forms of the growth. Toxicity should be considered as occurring, if specific oxygen uptake rates (SOUR) are severely depressed and protozoa and metazoa display sluggish activity or inactivity.

Poor preparation of the India ink/mixed liquor wet mount (Figure 34.17) also may result in false positives. Poor preparation can occur through the use of permanent India ink or blue india ink and the application of an insufficient quantity of india ink.

FILAMENTOUS ORGANISMS, ABSENCE OR PRESENCE

In a healthy activated sludge process, filamentous organisms should be present and easily observed during routine microscopic examinations of the mixed liquor (Table 34.21). If filamentous organisms are not easily observed, it may be that they are (1) transparent, short or growing mostly within the floc particles or (2) not capable of growing under the existing operational conditions. If filamentous organisms are not easily observed, a drop of methylene blue can be added to the wet mount. There are three operational conditions that prevent the growth of filamentous organisms. These conditions are (1) young sludge age or MCRT, (2) presence of complex wastes as the principle substrate, and (3) toxicity.

TABLE 34.21   Microscopic Set-up for the Presence or Absence of Filamentous Organisms

Slide Preparation	Microscope	Power of Magnification
Wet mount	Phase contrast	100× total or 1000× total

TABLE 34.22   Relative Abundance Ratings for Filamentous Organisms

Relative Abundance Rating	Term	Description
0	"None"	Filamentous organisms not observed
1	"Insignificant"	Filamentous organisms present, but found in an occasional floc particle in very few fields of view
2	"Some"	Filamentous organisms present, but found only in some floc particles
3	"Common"	Filamentous organisms observed in most floc particles at low density (1–5 filamentous organisms per floc particle)
4	"Very common"	Filamentous organisms observed in most floc particles at medium density (6–20 filamentous organisms per floc particle)
5	"Abundant"	Filamentous organisms observed in most floc particles at high density (>20 filamentous organisms per floc particle)
6	"Excessive"	Filamentous organisms observed in most floc particles; filamentous organisms more abundant than floc particles; or filamentous organisms growing in large numbers in the bulk solution

TABLE 34.23   Microscopic Set-up for Evaluating the Relative Abundance of Filamentous Organisms

Slide Preparation	Microscope	Power of Magnification
Wet mount	Bright-field or phase-contrast	100× total

FILAMENTOUS ORGANISMS, RELATIVE ABUNDANCE

The relative abundance of filamentous organisms in an activated sludge process and its impact on settleability and loss of secondary solids are commonly evaluated on a scale of "0" to "6", with "0" being "none" and "6" being "excessive" (Table 34.22). Ratings of "0," "1", and "2" do not adversely affect settleability of secondary solids. A rating of "3" typically does not adversely affect settleability of secondary solids. Ratings of "4," "5", and "6" usually do adversely affect settleability of secondary solids. The relative abundance of filamentous organisms can be observed with the bright-field or phase contrast microscope (Table 34.23). A drop of methylene blue may be added to the wet mount to more easily observe the filamentous organisms under bright-field microscopy.

TABLE 34.24   Location of Filamentous Organisms

Free-floating in the bulk solution
Extending into the bulk solutionfrom the perimeter of the floc particle
Mostly within the floc particle

TABLE 34.25   Microscopic Set-up for Determining the Location of Filamentous Organisms

Slide Preparation	Microscope	Power of Magnification
Wet mount	Bright-field or phase contrast	100× total

FILAMENTOUS ORGANISMS, LOCATION

Each filamentous organism typically grows in a specific location or locations in the mixed liquor (Table 34.24). The filamentous organism *Beggiatoa* may be found free-floating in the bulk solution. The filamentous organism *Sphaerotilus natans* typically is found extending into the bulk solution from the perimeter of the floc particle and occasionally free-floating in the bulk solution, while the filamentous organism type 0092 usually is found growing mostly within the floc particles. Chapter 31, "Filamentous Organisms", provides tables for major filamentous organisms that identify their location in mixed liquor.

The location of the filamentous organism is a characteristic that may be used to identify the organism to name or type number and may serve as a bioindicator of an unhealthy system. For example, filamentous organisms that typically grow in the floc particle or extend into the bulk solution from the perimeter of the floc particle may indicate the presence of shearing action, cell bursting agents, or surfactants, if they are found mostly free-floating in the bulk solution. Methylene blue may be added to the wet mount of mixed liquor to better observe the location of the filamentous organisms (Table 34.25).

FILAMENTOUS ORGANISMS, IDENTIFICATION

The identification of filamentous organisms to name or type number and their correlation to specific operational growth conditions is achieved through the identification of specific morphological or structural features, responses to specific stains, and the use of an appropriate taxonomic key. Bright-field microscopy and phase contrast microscopy are required to identify most filamentous organisms (Table 34.26). The identification of filamentous organisms is reviewed in Chapter 31, "Filamentous Organisms."

FILAMENTOUS ORGANISMS, DOMINANT AND RECESSIVE

Filamentous organisms that are rated "4," "5", or "6" with respect to their relative abundance are dominant, while those filamentous organisms that are rated "1," "2",

TABLE 34.26 Morphological Characteristics, Staining Responses, and "S" Test to be Determined by Bright-field and/or Phase Contrast Microscopy

Feature	Microscope
Attached growth	Bright-field or phase contrast
Branching	Bright-field or phase contrast
Cell shape	Bright-field or phase contrast
Cell size, μm	Bright-field or phase contrast
Constrictions	Bright-field or phase contrast
Crosswalls (septae)	Bright-field or phase contrast
Filament location	Bright-field or phase contrast
Filament shape	Bright-field or phase contrast
Filament size, μm	Bright-field or phase contrast
Motility	Bright-field or phase contrast
Sheath	Phase contrast
Sulfur granules	Phase contrast
Gram stain	Bright-field
Neisser stain	Bright-field
PHB stain	Bright-field
"S" test	Phase contrast
Sheath stain	Phase contrast

TABLE 34.27 Ratings for Interfloc Bridging and Open Floc Formation

Rating	Description
Insignificant	Interfloc bridging and/or open floc formation observed in few fields of view
Significant	Interfloc bridging and/or open floc formation observed in most fields of view

TABLE 34.28 Microscopic Set-up for Evaluating Interfloc Bridging and Open Floc Formation

Slide Preparation	Microscope	Power of Magnification
Wet mount	Bright-field or phase contrast	100× total

or "3" are recessive. Dominant filamentous organisms should be identified to name or type number and their growth correlated to the appropriate stimulatory operational conditions.

FLOC PARTICLE, INTERFLOC BRIDGING, AND OPEN FLOC FORMATION

There are five floc particle structures with respect to the growth or lack of growth of filamentous organisms. These structures are pin floc (pinpoint floc), ideal floc, filamentous bulking, interfloc bridging, and open floc formation (diffused floc). In addition to undesired numbers of filamentous organisms, interfloc bridging and open floc formation contribute significantly to settleability problems in the secondary clarifier (Table 34.27). Interfloc bridging and open floc formation can be observed with the bright-field or phase contrast microscope (Table 34.28).

Pin floc is the growth of floc particles at a young sludge age and the absence of filamentous organisms (Figure 34.18). Pin floc is small in size and spherical in shape. Pin floc typically is white or mostly white in color.

Ideal floc is the growth of floc particles with balanced growth between floc bacteria and filamentous organisms (Figure 34.19). The bacteria in the floc particle are grouped in one large mass and one to five filamentous organisms extend into the bulk solution from the perimeter of the floc particle. Ideal floc usually is medium or large in size and irregular in shape. Ideal floc typically is golden-brown in color.

Filamentous bulking (Figure 34.20) is the presence of undesired filamentous organism growth. The growth may be rated as "4," "5", or "6." Floc particles usually are medium and large in size and irregular in shape. Floc particles are typically golden-brown in color.

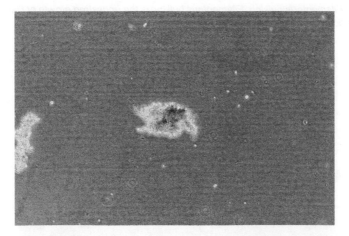

Figure 34.18 *Pin floc. Pin floc is white or mostly white, small (<150 μm), and spherical in shape. No or little filamentous organism growth is present.*

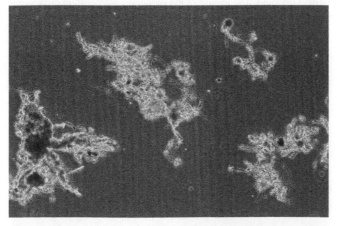

Figure 34.19 *Ideal floc. Ideal floc is mostly golden-brown, medium (150–500 μm) or large (>500 μm), and irregular in shape. Ideal floc has balanced growth between floc bacteria and filamentous organisms.*

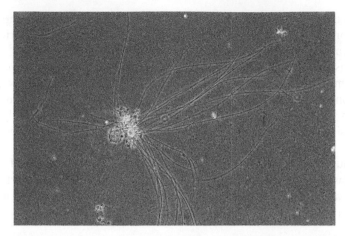

Figure 34.20 *Filamentous bulking. Filamentous bulking is the undesired growth of filamentous organisms in floc particles.*

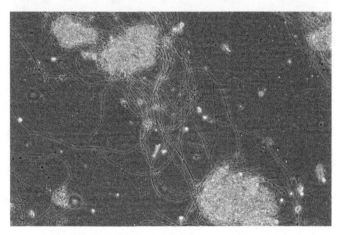

Figure 34.21 *Interfloc bridging. The joining of the extended filamentous organisms in the bulk solution from the perimeter of two or more floc particles is interfloc bridging.*

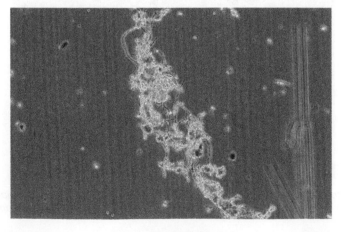

Figure 34.22 *Open floc formation. The scattering of floc bacteria in many small groups along the lengths of the filamentous organisms is open floc formation.*

Interfloc bridging is the joining in the bulk solution of the extended filamentous organisms from the perimeter of two or more floc particles (Figure 34.21). Floc particles usually are medium and large in size and irregular in shape. Floc particles are typically golden-brown in color.

Open floc formation is the scattering of the floc bacteria in many small groups along the lengths of the filamentous organisms within the floc particle (Figure 34.22). Floc particles usually are medium and large in size and irregular in shape. Floc particles are typically golden-brown in color.

35

Worksheets

There are four worksheets that may be used to describe or characterize the mixed liquor biota. The worksheets include the following data: (1) sample collection information, (2) bulk solution characterization, (3) floc particle characterization, (4) filamentous organism profile and identification, (5) protozoan count and profile, and (6) metazoan count and profile. Worksheet 35.1 reviews bulk solution, floc particles, and filamentous organisms. Worksheet 35.2 reviews protozoa and metazoa. In addition to these worksheets, a worksheet should be used to determine whether the microscopic observations are acceptable or unacceptable. Worksheet 35.3 provides this information. Worksheet 35.4 provides the total organism count.

Sample Collection Information

Sample collection information that should be provided on all worksheets includes the following:

- Location of the sample
- Date and time of the sample
- Chemicals used at the location
- Name of the individual who collected the sample
- Date and time of the microscopic examination
- Name of the microscopist

Microscopic Examination of the Activated Sludge Process, by Michael H. Gerardi
Copyright © 2008 John Wiley & Sons, Inc.

WORKSHEET 35.1   Microscopic Examination of Bulk Solution, Floc Particles, and Filamentous Organisms

Name of the wastewater treatment plant	
Location of sample	
Date of sample	
Chemicals used at the location	
Name of the individual collecting the sample	
Date of the microscopic examination	
Microscopist	
Bulk Solution	*Observation or Rating*
Particulate material (insignificant, significant)	
Dispersed growth (insignificant, significant, excessive)	
Floc Particles	*Observation or Rating*
Shape (irregular, oval, spherical)	
Size (small, medium, large)	
Ranges in sizes (μm)	
Color (golden-brown, light or white)	
Strength (firm, weak)	
Interfloc bridging (insignificant, significant)	
Open floc formation (insignificant, significant)	
India ink reverse stain (negative, positive)	
Zoogloeal growth (insignificant, significant)	
Zoogloeal growth (amorphous, dendritic)	
Filamentous Organisms	*Observation or Rating*
Relative abundance (0, 1, 2, 3, 4, 5, 6)	
Location (extending, free-floating, within floc particles)	
Ranges in lengths (μm)	
Dominant filamentous organisms	
Filamentous organism #1	
Filamentous organism #2	
Filamentous organism #3	
Recessive filamentous organisms	

WORKSHEET 35.1

Bulk Solution Characterization

Characterization of the bulk solution along with appropriate ratings that should be provided on the worksheet include the following:

- Particulate material
  - Insignificant
  - Significant
- Dispersed growth
  - Insignificant
  - Significant
  - Excessive

Floc Particle Characterization

Characterization of the floc particles along with appropriate ratings that should be provided on the worksheet include the following:

- Floc particle shape, dominant
  - Irregular
  - Oval
  - Spherical
- Floc particle size, dominant
  - Small (<150 μm)
  - Medium (150–500 μm)
  - Large (>500 μm)
- Range in sizes (μm)
- Floc particle color, dominant
  - Golden-brown
  - Light or white
- Strength, dominant
  - Firm
  - Weak
- Filamentous organism and floc particle structure
  - Interfloc bridging
    - Insignificant
    - Significant
  - Open floc formation
    - Insignificant
    - Significant
- India ink reverse stain (nutrient deficiency)
  - India ink reverse stain negative (no nutrient deficiency)
  - India ink reverse stain positive (probable nutrient deficiency)
- Zoogloeal growth (viscous floc)
  - Amorphous or globular
    - Insignificant
    - Significant
  - Dendritic or "fingerlike"
    - Insignificant
    - Significant

Filamentous Organism Profile and Identification

Characterization of the filamentous organisms in the mixed liquor biota is based on the relative abundance of the filamentous organisms, their locations and range in lengths, and identification of the filamentous organisms to scientific name such as

Haliscomenobacter hydrossis or type number such as type 1701. Characterization of the filamentous organisms along with appropriate ratings that should be provided on the worksheet include the following:

- Relative abundance rating of all filamentous organisms and relative abundance rating for each filamentous organism
  - "0"
  - "1"
  - "2"
  - "3"
  - "4"
  - "5"
  - "6"
- Ranges in lengths of most filamentous organisms, for example,
  - <50 μm, 100–200 μm, and >400 μm
  - <100 μm and >500 μm
- Location of filamentous organisms
  - Extending from the perimeter into the bulk solution
  - Free-floating in the bulk solution
  - Within the floc particles
- Identification (name or type number) of filamentous organisms in order of rank or dominance
  - Filamentous organism #1: _____
  - Filamentous organism #2: _____
  - Filamentous organism #3: _____

WORKSHEET 35.2

Protozoan Count and Profile

Characterization of the protozoan community is based on the relative abundance or count (protozoa per milliliter), dominant protozoan groups, commonly observed genera or species in the dominant groups, protozoan activity, and protozoan structure. Characterization of the protozoan community along with appropriate ratings that should be provided on the worksheet include the following:

- Count or number of protozoa per milliliter
- Profile or composition
  - % Amoebae
  - % Flagellates
  - % Free-swimming ciliates
  - % Crawling ciliates
  - % Stalked ciliates

WORKSHEET 35.2   Microscopic Examination of the Protozoan Community and Metazoan Community

Name of the wastewater treatment plant	
Location of sample	
Date of sample	
Chemicals used at the location	
Name of the individual collecting the sample	
Date of the microscopic examination	
Microscopist	
Protozoan Community	*Observation or Rating*
Count or number of protozoa per milliliter	
Profile % Amoebae	
% Flagellates	
% Free-swimming ciliates	
% Crawling ciliates	
% Stalked ciliates	
Commonly observed genera or species in dominant groups	
Activity (acceptable, not acceptable)	
% Stalked ciliates in free-swimming mode	
% Stalked ciliates forming gas bubbles	
% Stalked ciliates that are sheared	
Metazoan Community	*Observation or Rating*
Count (number of rotifers and free-living nematodes/mL)	
Activity (acceptable, not acceptable)	
% Rotifers and free-living nematodes that are dispersed	
Spirochetes and Tetrads	*Observation or Rating*
Spirochetes (insignificant, significant)	
Tetrads (insignificant, significant)	
Tetrads (adsorbed to floc particles, free-floating)	
Additional Organisms	*Observation or Rating*
Algae	
Bloodworms and bristleworms	
Copepods and cyclops	
Gastrotriches	
Sludge worms	
Water bears	
Water fleas	

- Commonly observed genera or species in dominant groups
- Activity
  - Acceptable
  - Not acceptable
- Structure
  - % Stalked ciliates in free-swimming mode
  - % Stalked ciliates forming gas bubbles
  - % Stalked ciliates that are sheared

Metazoan Count and Profile

Characterization of the metazoan community is based on the numbers, activity, and structure of rotifers and free-living nematodes. Characterization of the metazoan community along with appropriate ratings that should be provided on the worksheet include the following:

- Count or number of rotifers and free-living nematodes per milliliter
- Activity
  - Acceptable
  - Not acceptable
- Percent of rotifers and free-living nematodes that are dispersed

Additional components of the mixed liquor biota that may be important in troubleshooting the activated sludge process include the presence of significant numbers of spirochetes and tetrads. Spirochetes are found free swimming in the bulk solution. Tetrads are found in the bulk solution and mostly adsorbed to floc particles. A notation of the presence, activity, and structure of algae, bloodworms, bristleworms, copepods, cyclops, gastrotriches, sludge worms, water bears, water fleas, and other organisms may be included in the worksheet.

In nearly all operational conditions including the steady-state condition, microscopic examination of the bulk solution, floc particles, and filamentous organisms is routinely performed or needed (Worksheet 35.1). Microscopic examination of the protozoan community and the metazoan community are performed occasionally during the steady-state condition and recommended during an adverse operational condition to obtain more troubleshooting information (Worksheet 35.2).

The microscopic observations recorded on Worksheet 35.1 and Worksheet 35.2 should be compared to those observations that are typically observed during a healthy, steady-state operational condition (Worksheet 35.3). The observations should be recorded as "acceptable" or "not acceptable." "Not acceptable" ratings require troubleshooting and corrective efforts to return the activated sludge process to its "acceptable," steady-state operational condition.

WORKSHEET 35.4

Total Organism Count

A total organism count of the mixed liquor consists of the enumeration of protozoa, rotifers, and free-living nematodes in a milliliter of sample. This technique correlates operational conditions and process changes to organism count. Because of the relatively long generation times of rotifers and free-living nematodes, a total organism count should only be performed on an activated sludge process with a MCRT ≥28 days.

The enumeration or counting of organisms is performed by preparing a wet mount that contains approximately 0.05 mL of sample and examining the wet mount at 100× total magnification with bright-field or phase contrast microscopy. Phase contrast microscopy is preferred. A 22×22 mm coverslip must be used in preparing this wet mount.

WORKSHEET 35.3   Microscopic Examination, Observations, and Ratings

Name of the wastewater treatment plant	
Location of sample	
Date of sample	
Chemicals used at the location	
Name of the individual collecting the sample	
Date of the microscopic examination	
Microscopist	

Observation	Typical Rating at Steady State	Current Rating	
		Acceptable	Not Acceptable
Bulk Solution			
Particulate material			
Dispersed growth			
Floc Particles			
Shape			
Size			
Range in sizes			
Color			
Strength			
Interfloc bridging			
Open floc formation			
India ink stain			
Zoogloeal growth			
Filaments			
Abundance			
Location			
Range in lengths			
Dominant filaments			
Recessive filaments			
Protozoa			
Count			
Dominant groups			
Dominant genera			
Activity			
Structure			
Metazoa			
Count			
Activity			
Structure			

WORKSHEET 35.4   Total Organism Count

Organism	Scan	Count	Average of 4 Counts
Protozoa	#1		
	#2		
	#3		
	#4		
Rotifers	#1		
	#2		
	#3		
	#4		
Free-living nematodes	#1		
	#2		
	#3		
	#4		
Total organism count			

The protozoa, rotifers, and free-living nematodes are counted by scanning the entire coverslip area. The scanning results of four separate wet mounts are averaged, and the average count is multiplied by 20 to obtain the number of organisms per milliliter of mixed liquor (Worksheet 35.4).

When a large number of organisms are present in the mixed liquor, the number of organisms present per field of view (100× total magnification) is counted for 10 fields of view selected at random. The average number for these 10 fields of view is multiplied by 300 (approximate number of fields of view under a 22 × 22 mm coverslip at 100× total magnification) to obtain the number of organisms under the coverslip.

36

Report of Microscopic Examination

Although worksheets and rating tables are excellent formats for recording and evaluating microscopic observations and quickly noting atypical conditions in the mixed liquor biota food chain for operational personnel, they may be awkward in providing operational information to nonoperational personnel. Therefore, an appropriate descriptive report of the microscopic observations and their correlation to operational conditions is beneficial. The sample report that is provided in this chapter provides a guide for the preparation of reports for the microscopic examination of wastewater samples.

***

Microscopic Examination
City of Smithville, Pennsylvania
Conventional Activated Sludge Process with Significant Industrial Discharge

Sampling Date: 21 February 2007
Microscopic Examination Date: 21 February 2007
Examination Performed by: J. P. Gerardi

Samples Examined:
Mixed Liquor, Aeration Basin #1, Mixed Liquor, Aeration Basin #2, Foam, Aeration Basin #1

Mixed Liquor, Aeration Basin #1
Floc Characterization and Filamentous Organisms

Although the range in sizes of the floc particles was approximately 50 to 1500 μm, most floc particles were medium (150 to 500 μm) and large (>500 μm) in size. Due to the presence of a copious growth of filamentous organisms, most floc particles were irregular in shape and golden-brown in color. The range in sizes and color of the floc particles are typical of an active and mature mixed liquor, and the presence of irregular floc particles also is typical when a large population of filamentous organisms is present. However, the size of the population of filamentous organisms is undesired.

As revealed through phase contrast microscopy and methylene blue staining, most floc particles were firm in structure. These particles possessed floc bacteria that were tightly adjoined. Firm structure in floc particles is highly desired.

Significant interfloc bridging and significant open floc formation were observed. Interfloc bridging is the joining in the bulk solution of the extended filamentous organisms from two or more floc particles. Open floc formation is the scattering of the floc bacteria in many small groups along the lengths of the filamentous organisms. Significant interfloc bridging and significant open floc formation adversely affect solids settleability and compaction.

Numerous filamentous organisms were observed extending into the bulk solution from the perimeter of the floc particles and within the floc particles. The filamentous organisms within the floc particles were not observed until Gram staining of a smear of mixed liquor was performed. According to appropriate taxonomic keys the filamentous organisms were observed in expected locations. An occasional free-floating filamentous organism was observed. The ranges in lengths of most filamentous organisms were <150 μm and >400 μm.

The relative abundance of the filamentous organisms was rated as "5" on a scale of "0" to "6," with "0" being "none" and "6" being "excessive." A rating of "5" is "abundant" or "filamentous organisms observed in most floc particles at high density, for example, >20 filamentous organisms in most floc particles." At this relative abundance rating the filamentous organisms would adversely affect solids settleability and compaction.

There were three significant filamentous organisms. A "significant" filamentous organism is one that has a relative abundance rating of "4" or "higher." The significant filamentous organisms, their dominance or rank and their relative abundance ratings are listed in Table 1.

Microthrix parvicella is a foam-producing filamentous organism. Foam typical of this organism is viscous and chocolate-brown. *Microthrix parvicella* usually is found in higher density in the foam than mixed liquor.

Operational factors associated with the rapid and undesired growth of *Thiothrix*, *Microthrix parvicella* and *Sphaerotilus natans* are listed in Table 2.

The bulk solution contained much dispersed growth and much particulate material. The relative abundance of dispersed growth was rated as "excessive," and the

TABLE 1   Significant Filamentous Organisms, Mixed Liquor, Aeration Basin #1

Filamentous Organism	Rank	Relative Abundance
Thiothrix	1	"5"
Microthrix parvicella	2	"4"
Sphaerotilus natans	3	"4"

TABLE 2 Operational Conditions Associated with the Undesired Growth of Observed Significant Filamentous Organisms

Operational Condition	Filamentous Organism		
	Thiothrix	Microthrix	Sphaerotilus
High MCRT (>10 days)		X	
Fats, oils, and grease		X	
High pH (>7.5)		X	
Low DO and high MCRT		X	
Low DO and low/moderate MCRT			X
Low F/M (<0.05)		X	
Low nitrogen or phosphorus	X		X
Organic acids	X		
Readily degradable cBOD	X		X
Septicity/sulfides	X		
Slowly degradable cBOD		X	
Cold wastewater temperature		X	
Warm wastewater temperature			X

TABLE 3 Operational Factors Associated with the Interruption of Floc Formation

Operational Factor	Description or Example
Cell bursting agent/surfactant	Lauryl sulfate
Foam production	Foam-producing filamentous organisms
Low dissolved oxygen concentration	<1 mg/L for 10 consecutive hours
Low pH or high pH	<6.5 or >8.0
Nutrient deficiency	Usually nitrogen or phosphorus
Slug discharge of soluble cBOD	3× normal quantity of soluble cBOD
Toxicity	RAS chlorination
Undesired filamentous organism growth	Relative abundance rating >"3"
Young sludge age	<3 days MCRT
Zoogloeal growth	Rapid floc-forming bacterial growth

relative abundance of particulate material was rated as "significant." The presence of excessive dispersed growth and significant particulate material is indicative of the interruption of floc formation and may be associated with one or more of the operational factors listed in Table 3.

Significant dendritic or "fingerlike" zoogloeal growth (viscous floc) was observed. Zoogloeal growth is the rapid and undesired proliferation of floc-forming bacteria. Over time, the growth results in the production of weak and buoyant floc particles. The growth also may be associated with the production of billowy white foam. Operational conditions responsible for undesired zoogloeal growth include (1) high MCRT, (2) long HRT, (3) nutrient deficiency, (4) high F/M, and (5) septicity or fermentation upstream of the aeration tank.

Most floc particles tested positively to the india ink reverse stain. The positive test results are suggestive of a probable nutrient deficiency at the time of sampling. Nutrients typically found deficient in an activated sludge process are nitrogen and phosphorus.

Protozoan Count and Profile

The count or relative abundance of the protozoan community was 1400 per milliliter. This count is significantly less than the range of values for the protozoan community at a steady-state, healthy condition from previous examinations. Although the community was dominated by the higher life-forms, crawling ciliates and stalked ciliates, the ciliates were dominated by *Aspidisca costata* (crawling ciliate) and *Vorticella alba* (stalk ciliate), which are indicative of an intermediate sludge that produces an acceptable but not highly "polished," mixed liquor effluent. The profile or percent composition of the protozoan community is provided in Table 4.

TABLE 4 Protozoan Profile, Mixed Liquor, Aeration Basin #1

Protozoan Group	Composition in the Community
Amoebae	4%
Flagellates	19%
Free-swimming ciliates	1%
Crawling ciliates	21%
Stalked ciliates	55%

TABLE 5 Microscopic Observations for Mixed Liquors, Aeration Basins #1 and #2

Microscopic Observation	Mixed Liquor, Aeration Basin	
	#1	#2
Floc particles, range in sizes (μm)	50 to 1500	40 to 1200
Floc particles, dominant sizes	Medium and large	Medium and large
Floc particles, dominant shape	Irregular	Irregular
Floc particles, strength	Firm	Firm
Interfloc bridging	Significant	Significant
Open floc formation	Significant	Significant
Filamentous organisms, location	Extending/within floc	Extending/within floc
Filamentous organisms, lengths (μm)	<150 and >500	<150 and >500
Filamentous organisms, abundance	"5"	"5"
Filamentous organism, #1	*Thiothrix*	*Thiothrix*
Filamentous organism, #2	*Microthrix*	*Microthrix*
Filamentous organism, #3	*Sphaerotilus*	*Sphaerotilus*
Dispersed growth	Excessive	Significant
Particulate material	Significant	Significant
Zoogloeal growth	Significant	Significant
India ink reverse stain	Positive	Positive
Protozoan count	1400 per milliliter	1200 per milliliter
% Amoebae	4	5
% Flagellates	19	18
% Free-swimming ciliates	1	3
% Crawling ciliates	21	24
% Stalked ciliates	55	50
Dominant crawling ciliate	*Aspidisca costata*	*Aspidisca costata*
Dominant stalked ciliate	*Vorticella alba*	*Vorticella alba*
% Free-swimming stalked ciliates	30%	18%
Protozoan activity	Sluggish	Sluggish
Metazoan activity	Sluggish	Sluggish
Dispersed metazoa	Significant	Significant

Approximately 30% of the stalked ciliates were observed to be swimming freely in the bulk solution. In addition, as noted by the slow beating of cilia and slow "springing" action of the contractile filament in some of the stalked ciliates, many of the stalked ciliates were sluggish in activity. The dominant life-forms for protozoa, free-swimming stalked ciliates, and sluggish activity of the protozoan community may be indicative of (1) low dissolved oxygen concentration and (2) presence of inhibitory or toxic wastes including surfactants.

Metazoan Count and Profile

In addition to the protozoan community, a relatively small and sluggish community of metazoa (rotifers and free-living nematodes) was observed. The count or relative abundance of the community was <100 per milliliter. Nearly all observed rotifers and free-living nematodes were sluggish or inactive, and many of the metazoa were dispersed. This community also be indicative of the presence of (1) low dissolved oxygen concentration and (2) presence of inhibitory or toxic wastes including surfactants.

Mixed Liquor, Aeration Basin #2
Floc Characterization, Filamentous Organisms, Protozoan Count and Profile, Metazoan Count and Profile

Microscopic observations for floc characterization, filamentous organisms, protozoa and metazoa, in Mixed Liquor, Aeration Basin #2 are listed in Table 5 along with the microscopic observations for Mixed Liquor, Aeration Basin #1.

Foam, Aeration Basin #1
Characterization

Wet mounts and Gram-stained and Neisser-stained smears of foam revealed the presence of a significant growth of *Microthrix parvicella*. This filamentous organism was found in floc particles, extending from the perimeter of floc particles and between floc particles. The filamentous organism was observed in greater density in foam than mixed liquor. The relative abundance of *Microthrix parvicella* in foam and the texture and color (viscous and chocolate-brown) of foam indicate that *Microthrix parvicella* is responsible for foam production.

References

Arnz, P., E. Arnold, and P. A. Wilderer. 2000. Enhanced biological phosphorus removal in a semi full-scale SBR. *Wat. Sci. Technol.* 43.

Calaway, W. T. 1968. The metazoa of waste treatment processes—Rotifers. *J. Water Polln. Control Fed.* 40.

Calaway, W. T. and J. A. Lackey. 1962. *Waste Treatment Protozoa; Flagellata. Florida Engineering Series No. 3.* University of Florida, Gainesville, FL.

Carrias, J., C. Amblard, and G. Bourdier. 1996. Protistan bacterivory in an oligomesotrophhic lake; importance of ciliates and flagellates. *Microb. Ecol.* 31.

Curds, C. R. 1992. *Protozoa in the Water Industry.* Cambridge University Press. Cambridge, U.K.

Curds, C. R. 1969. *An Illustrated Key to the British Freshwater Ciliated Protozoa Commonly Found in Activated Sludge.* Tech. Paper No. 12. Water Pollution Resources, Ministry of Technology. London, U.K.

Curds, C. R. and A. Cockburn. 1973. The role of protozoa in the activated sludge process. *Amer. Zool.* 13.

Eikelboom, D. H. and H. J. J. van Buijsen. 1981. *Microscopic Sludge Investigation Manual.* TNO Res. Inst. Environ. Hyg. Netherlands.

Esteban, G., C. Tellez, and L. M. Bautista. 1990. Effects of habitat quality on ciliated protozoa communities in sewage treatment plants. *Envir. Technol.* 12.

Fenchel, T. 1987. *Ecology of Protozoa; The Biology of Free-living Phagotrophic Protists.* Springer-Verlag. New York.

Fernandez-Leborans, G. and A. Novillo. 1996. Protozoan communities and contamination of several fluvial systems. *Wat. Envir. Res.* 68.

Genoveve, E., C. Tellez, and L. M. Batista. 1991. Dynamics of ciliated protozoa communities in activated-sludge process. *Wat. Res.* 25.

Gerardi, M. H. 1987. An operator's guide to free-living nematodes in wastewater treatment. *Pub. Works.* 118.

Gerardi, M. H. 1986. An operator's guide to protozoa. *Pub. Works.* 7.

Gerardi, M. H. 1984. An operator's guide to rotifers and wastewater treatment processes. *Pub. Works.* 11.

Haller, E. J. 1991. Bugs in activated sludge: how do they work? *Water/Eng. & Management.* 8.

Hobson, T. 1986. Process control fundamentals, part VI; microscopic examination of activated sludge & control of aeration rates. *Operations Forum.* Water Pollution Control Federation. 11.

Jenkins, D., M. G. Richard, and G. T. Daigger. 1993. *Manual on the Causes and Control of Activated Sludge Bulking and Foaming,* 2nd Edition. Lewis Publishers, Chelsea, MI.

Laybourn-Parry, J. 1992. *Protozoan Plankton Ecology.* Chapman & Hall, New York.

Madoni, P., D. Davoli, G. Gessica, and L. Vescovi. 1996. Toxic effect of heavy metals on the activated sludge protozoan community. *Wat. Res.* 30.

Mitchell, R., Editor. 1972. *Water Pollution Microbiology.* Wiley-Interscience, New York.

Mudrack K. and S. Kunst. 1986. *Biology of Sewage Treatment and Water Pollution Control.* John Wiley & Sons, New York.

Ratsak, C. H., K. A. Maarsen, and S. A. L. M. Kooijman. 1996. Effects of protozoa on carbon mineralization in activated sludge. *Wat. Res.* 30.

Richard, M. 1989. *Activated Sludge Microbiology.* Water Pollution Control Federation, Alexandria, VA.

Salvado, H. and M. P. Gracia. 1993. Determination of organic loading rate of activated sludge plants based on protozoan analysis. *Wat. Res.* 27.

Salvado, H. M. P. Gracia, and J. M. Amigo. 1993. Capability of ciliated protozoa as indicators of effluent quality of activated sludge plants. *Wat. Res.* 29.

Sigee, D. C. 2005. *Freshwater Microbiology; Biodiversity and Dynamic Interactions of Microorganisms in the Aquatic Environment.* John Wiley & Sons, Ltd., Chichester, UK.

Spellman, F. R. 1997. *Microbiology for Water/Wastewater Operators.* Technomic Publishing Co., Lancaster, PA.

Spellman, F. R. and J. E. Drinan. 2001. *Stream Ecology & Self-Purification; An Introduction.* Technomic Publishing Co., Lancaster, PA.

Williams, T. M. and R. F. Unz. 1983. Environmental distribution of Zoogloea strains. *Water Res.* 17.

Abbreviations and Acronyms

BOD	Biochemical oxygen demand
°C	Degrees Celsius
cBOD	Carbonaceous biochemical oxygen demand
DO	Dissolved oxygen
F/M	Food-to-microorganism
FOG	Fats, oils and grease
HRT	Hydraulic retention time
I/I	Inflow and infiltration
MCRT	Mean cell residence time
mL	Milliliter
MLVSS	Mixed liquor volatile suspended solids
mm	Millimeter
mV	Millivolt
nBOD	Nitrogenous biochemical oxygen demand
nm	Nanometer
ORP	Oxidation-reduction potential
PAO	Phosphorus accumulating organism
RAS	Return activated sludge
SOUR	Specific oxygen uptake rate
SRB	Sulfate reducing bacteria
TFO	Tetrad forming organisms

TSS	Total suspended solids
WET	Whole effluent toxicity
w/v	Weight-to-volume
β	Beta
μm	Micron

Chemical Compounds

CO_2	Carbon dioxide
H_2S	Hydrogen sulfide
$HgCl_2$	Mercuric chloride
HPO_4^{2-}	Orthophosphate (reactive phosphorus)
$H_2PO_4^-$	Orthophosphate (reactive phosphorus)
N_2	Molecular nitrogen
Na_2S	Sodium sulfide
$NiSO_4$	Nickel sulfate
NH_4^+	Ionized ammonia
NH_4^+-N	Ammonical-nitrogen
NO_2^-	Nitrite
NO_3^-	Nitrate
NO_3^--N	Nitrate-nitrogen
O_2	Dissolved oxygen
SO_4^{2-}	Sulfate

Glossary

acetogene   an organism that produces acetate through fermentative processes

acute   short-term

aerobe   an organism that can live and grow only in the presence of free molecular oxygen

agglutinate   cement together by sticky substances

bacillus   rod-shaped

binary fission   division of a cell into two daughter cells

bioaugmentation   the addition of commercially prepared bacterial and fungal cultures

biochemical   chemical reaction occurring inside living cells

biomass   the living mass of an animal and/or plant population

biota   a living component

budding   the production of daughter cells in the form of rounded outgrowths

carbonaceous   organic compound; a compound containing carbon and hydrogen

chitin   a skeletal material found in the majority of groups of invertebrates

chloroplast   plastid containing chlorophyll embedded in plant cells

chronic   long-term

cilia   short hair-line structures found on many protozoa and metazoa that are used for locomotion and food gathering

cirri   modified cilia found on crawling ciliated protozoa that are used to anchor the organism on floc particles

coccus   spherical shape

colloid   molecule with a large surface area that does not dissolved in water and does not settle in water

cytoplasm   aside from the nucleus, the gut content of a cell

dimorphic   two forms

enzyme   a proteinaceous molecule used by cells to increase the rate of a biochemical reaction; the enzyme is not consumed during the reaction

exocellular   outside the cell

exoskeleton   skeleton found outside the body

exponential growth   a stage of growth occurring in populations of unicellular microorganisms when the logarithm of the cell number increases linearly with time

facultative   able to live under different conditions

flagellum   whip-like structure found on most bacteria and some protozoa, used for locomotion and food gathering purposes

flocculation   the coalescence of finely divided precipitate and/or bacteria into large particles

habitat   where an organism lives

holozoic   devouring other organisms

hydrolytic   splitting of molecules with the use of water

inhibition   the stopping or deceleration of a metabolic process

intracellular   within the cell

metamorphosis   pronounced change of form and structure taking place within a comparatively short time

metazoa   a subkingdom of animals, comprising multicellular animals having two or more tissue layers

morphology   form or structure

nitrogenous   nitrogen-containing

nocardioform   group of truly branched, Gram-positive actinomycetes that are responsible for the production of viscous, chocolate-brown foam

oligochaete   a class of worms with relatively few chaetae, terrestrial and aquatic earthworms

organotroph   bacteria that degrade carbonaceous biochemical oxygen demand compounds

pathogen   organism that causes disease

polysaccharide   a group of complex carbohydrates such as starch and cellulose

saprobic   a measurement of the degree of pollution in a flowing body of water

saprophyte   any bacterium that breaks up dead animal and vegetable matter and does not cause disease in the animal or plant that it inhabits

saprozoic   devouring dead animal and vegetable matter

sessile   fixed and stationary

solitary   single or alone

spirillum   spiral shaped

thermophile   an organism that thrives at elevated temperatures

trichome   the thread of cells that together with a sheath makes of a filament

Index

Microscopic Examination of the Activated Sludge Process, by Michael H. Gerardi
Copyright © 2008 John Wiley & Sons, Inc.

243

Printed and bound by CPI Group (UK) Ltd, Croydon, CR0 4YY

27/10/2024

14580263-0004